DEMYSTIFYING THE AKASHA

Also by Ralph Abraham
(with Terence McKenna and Rupert Sheldrake)

Chaos, Creativity, and Cosmic Consciousness
The Evolutionary Mind

Also by Sisir Roy

**Statistical Geometry and Applications to Microphysics
and Cosmology**

The cover illustration, by Ralph Abraham and Peter Broadwell, is from one of the earliest computer simulations of the akasha. See Appendix 2.

DEMYSTIFYING THE AKASHA

Consciousness and the Quantum Vacuum

Ralph Abraham and Sisir Roy

Epigraph Books
Rhinebeck, New York

Demystifying the Akasha: Consciousness and the Quantum Vacuum. Copyright © 2010 by Ralph Abraham. All rights reserved. No part of this book may be used or reproduced in any manner without written permission from the publisher except in critical articles and reviews. For information contact:
Epigraph Publishing Service
27 Lamoree Road
Rhinebeck, New York 12572

Printed in the United States of America

Library of Congress Catalogue-in-Publication Data

Abraham, Ralph

> *Demystifying the Akasha: Consciousness and the Quantum Vacuum* by Ralph Abraham.

x + 211pp.
ISBN 978-0-9826441-5-7
1. Historical models. 2. Our models – 1936.

Library of Congress Control Number: 2010931786

Bulk purchase discounts for educational or promotional purposes are available. Contact the publisher for more information.

Epigraph Publishing Service
27 Lamoree Road
Rhinebeck, New York 12572
www.epigraphPS.com

Contents

Preface	viii
HISTORICAL INTRODUCTION	1
PART ONE, Historical Models	4
Ch. 1. Consciousness Models, Western	5
Ch. 2. Consciousness Models, Eastern,	25
Ch. 3. Atomism	41
Ch. 4. Quantum Theory and Consciousness	56
Ch. 5. Paranormal Research	68
PART TWO, Our Models	76
Ch. 6. The RR Model	78
Ch. 7. The AR Model	85
Ch. 8. The Consciousness Model	108
CONCLUSION	117
APPENDIX 1: Requardt and Roy (2001)	121
APPENDIX 2: Vibrations and Forms (2006)	161
APPENDIX 3: Technical Summary of AR (2006)	187
REFERENCES	190
Guide to Pronunciation	205
Glossary	207
Index	209

Analytical Contents

Preface

HISTORICAL INTRODUCTION

PART ONE, Historical Models

Ch. 1. Consciousness Models, Western
  1.1. Ancient Egypt, 2500-600 BCE
  1.2. Ancient Greece, 600 BCE 500 CE
  1.3. Middle Ages, 800-1400 CE
  1.4. Renaissance, 1400-1600 CE
  1.5. Baroque, 1600-1750 CE
  1.6. Modern Times, 1900 CE
  1.7. Maps of Consciousness

Ch. 2. Consciousness Models, Eastern
  2.1. Trika in Context
  2.2. Trika Creation Story
  2.3. Selected Concepts of Trika
  2.4. Akasa

Ch. 3. Atomism
  3.1. Ancient Greece
  3.2. Oriental and Islamic Views
  3.3. The Renaissance
  3.4. Modern Quantum Theory

Ch. 4. Quantum Theory and Consciousness
  4.1. Measurement Theory
  4.2. Consciousness and the Quantum Brain
  4.3. Consciousness and the Quantum Vacuum

Ch. 5. Paranormal Research
  5.1. Dean Radin
  5.2. Rupert Sheldrake
  5.3. Theory

PART TWO, Our Models

Ch. 6. The RR Model
    6.1. The QX Network
    6.2. The Condensation Process
    6.3. Afterword on Process Physics
Ch. 7. The AR Model
    7.1. An outline of the AR process
    7.2. The QX model
    7.3. The ST model
    7.4. The Spatial Organization
    7.5. Possible Implications
Ch. 8. The Consciousness Model
    8.1. The Mind/Body Problem
    8.2. The AR Model
    8.3. The Mind/Body Problem Resolved
    8.4. Summary

CONCLUSION

APPENDIX 1: Requardt and Roy, 2001)
APPENDIX 2: Vibrations and Forms, 2006
APPENDIX 3: Technical Summary of AR, 2006

BIBLIOGRAPHY

Guide to Pronunciation (Greek, Sanskrit)
Glossary (Greek, Sanskrit)

Index

Preface

From Ralph:

Since my first visit to India in 1972, I had been thinking about mathematical models for consciousness. Thanks to the West Bengal University of Technology (WBUT) and the Fulbright Program of the US Department of State, I was able to visit Kolkata for the month of January, 2006. Again in 2008 and 2009, with help from the Indian Statistical Institute, Kolkata (ISI), I was able to return for short visits. During all three visits I was lodged in the Ramakrishna Mission Institute of Culture (RMIC), which is a spiritual and intellectual haven. During my 2006 visit the RMIC hosted an International Conference on Science and Consciousness, to which I was happily invited. Here I discovered among scientists an openness to consciousness studies which is as common in India as it is rare in Europe and the US.

In one of Ervin Laszlo's books on consciousness and the quantum vacuum, I wrote in the Foreword, "When a great grand unified theory will appear it will very likely conform to the prophetic vision of Ervin Laslo."[1] This suggests "If and when", revealing my basic pessimism: Is, actually, the quantum vacuum field the basis of consciousness, as advertised in Laszlo's subtitle of this book, *Foundations of an Integral Science of Quantum, Cosmos, Life, and Consciousness*? For me, quantum mechanics is an incomplete theory, an hypothesis. So indeed, I never expected to be the coauthor of a book such as this one, based as it is on the mathematics of the quantum vacuum!

It happened like this. My 2006 stay Kolkata resulted in several invitations to speak in the local universities. One of these was from Professor Sisir Roy, who works on many frontiers of theoretical physics, including the foundations of quantum theory, to

---
[1](Laszlo, 2003)

speak in his department at the Indian Statistical Institute. This meeting with Sisir naturally led to a joint work on the mathematics of the quantum vacuum (see Chapter 7) having nothing to do with the science of consciousness, while simultaneously, the RMIC conference was ongoing. As we have meditation practice in common, it was not long before we adapted our model for the quantum vacuum to a model for consciousness (see Chapter 8).

I am deeply indebted to the Fulbright Program, WBUT, ISI, and the RMIC for support and encouragement in my part of this joint work with Sisir. And we are both very much indebted to Professor Debabrata SenSharma of the RMIC Indology Research Center for many kindnesses, including patient and extensive instructions in his specialty, Kashmiri Shaivism. Thanks also to our alpha readers: Paul Lee, Steve Rooke, and Aubrey MIller. Finally, my wife, Ray Gwyn Smith, and our dog, Shea, have been very supportive throughout the process.

Santa Cruz, June 6, 2010

From Sisir:

I have been working on Planck scale physics and discreteness vs continuum of spacetime since the beginning of the 21st century. The epistemological issues involved in this context attracted me towards Buddhist and Yoga philosophy. Buddhist scholar Vasuvandhu made a critical analysis of discreteness vs continuum, and the Hindu scholar Kanad propounded the concept of atom, many centuries ago. In this process of learning the various epistemological debates, I met Professor Ralph Abraham and his deep interest in these issues. So a fruitful collaboration began and led to this book. My wife Malabika Roy has been very supportive and helped me a lot in clarifying these epistemological debates among Buddhist and Hindu scholars.

Kolkata, May 14, 2009

HISTORICAL INTRODUCTION

What do we mean by the word consciousness? It is a difficult word, and occurs very frequently in this work. Besides its more common meaning of individual mental awareness, it may also include the personal unconscious system, and the collective mind, conscious and unconscious. It is this latter meaning that we generally intend by this word, sometimes called *cosmic consciousness*.

Models for consciousness mostly use the apparatus of mathematical physics: curved spaces, continuous fields, dynamical systems, and so on. The field metaphor for consciousness has a long history in India, where the *ākāśa* (akasha), or ether, is one of the five elements of the material world. In the medieval literature of Kashmiri Shaivism, the metaphor of *spanda*, or vibration, is fundamental to a model of consciousness with many *tattvas*, or levels or categories, and implies an awareness of the field concept. There is also an ancient awareness of the etheric field in the West, as for example in the *apeiron* of the Greek philosopher Anaximander (6th C. BCE).

In the West, the field model – popularized by Madame Blavastky (1877), Rudolph Steiner (1909), Teilhard de Chardin (1955), Fritjof Capra (1975), Itzhalk Bentov (1977), Rupert Sheldrake (1981), and others – has become widespread. Up to 1991, all of these developments have involved only continuous fields and their vibrations, like water waves, good vibrations, waves of consciousness, and the like. Meanwhile, after the revival of ancient atomistic thinking in the quantum revolution around 1900, quantum fields have entered the conversations of mystics as well as scientists, such as Fred Alan Wolf (1981), Amit Goswami (1986), and Ervin Laszlo (1987).

Quantum mechanics began as an experimental branch of atomic physics. The behavior of elementary particles, electrons or photons for example, became visible with the invention of the cloud chamber by Charles Wilson, a Scottish physicist, around 1900. The observation of pairs of elementary particles emerging out of (and disappearing into) nowhere by Carl Anderson in 1932 led to a new understanding of the vacuum as a field of energy, called the quantum vacuum, or equivalently, the zero point field. (See Chapter 3.) Specifically, it is the quantum vacuum that has come to the fore as a favorite metaphor in consciousness studies, and even identified explicitly with the akashic field in a series of books by Ervin Laszlo since 1987.

In this book we prepose a mathematical model for the quantum vacuum, originally due to Requardt and Roy (see Chapter 6), as a model for consciousness. Although we have taken this model from the physics of the quantum vacuum, we do not mean to suggest that the quantum vacuum is identical to the field of consciousness. Rather, we take a mathematical model for the quantum vacuum and repurpose it as a model for cosmic consciousness. We were attracted to this model for its potential to incorporate several effects.

First of all, we wanted our model to be compatible with the so-called paranormal phenomena of individual psychology – telepathy, clairvoyance, precognition, and so on – as the tendency of science to reject the extensive research results on these effects is partly due to the incompatibility of older models for consciousness. This historical incompatibility is particularly troublesome in the case of the time-dislocation phenomena – precognition, presentiment, and retrocausation – as we explain in Chapter 5.

Secondly, we wanted to contribute a new insight to the infamous mind/body problem – following Kashmiri Shaivism and a suggestion of the late Maurice Merleau-Ponty (1968) – in placing consciousness external to physical spacetime. It is difficult for us, as concrete thinkers, to imagine anything (even God) outside of space. Where would that be? Well, mathematics is one way to imagine this. Similarly, math gives us a way to imagine something that is outside of time. Not in space. Not in time. That is our destination.

Thirdly, we wanted to incorporate the entanglement of consciousness. This is a new way to think about paranormal phenomena such as telepathy that is inspired by the new discovery of *nonlocality* in quantum physics. This is the experimental observation of an effect predicted by Einstein, in which two elementary particles exchange information without utilizing any known field. The quantum fields of the two particles are said to be *entangled*. By analogy, two minds may be said to be entangled if they can communicate without resorting to any physical means of communication.

Finally, we wanted to build upon the extensive philosophy of the West (the tradition of the soul and spirit from Plato to Jung, see Chapter 1) and the East (the tattvas and spanda of Kashmiri Shaivism in the Sanskrit tradition, see Chapter 2). By standing on the shoulders of these very evolved theories of consciousness we hope to be understandable. Our model, incorporating these several effects is fundamentally digital, that is, not in any analog or continuous context. Thus it falls into the established category of digital philosophy (see Chapter 3).

We discuss the background data (philosophy, quantum concepts, and parapsychology) in Part One. Then in Part Two, we present our model step-by-step. All these threads are brought together in the Conclusion. Our main discussion, on the construction of continuum spacetime from the discrete akasa, is meant to be potentially compatible with new thinking in physics, such as process physics, quantum theory, and general relativity.

"The Sanskrit word *akasa* (ether or space) derives from *a* (towards) and *kasa* (to be visible, to appear). Akasa is the subtle "background" against which everything in the material universe becomes perceptible."[2]

---

[2](Yogananda, 2007; p. 867)

PART ONE: Historical Models

We are concerned with models for consciousness. The word *model* has many meanings. In the most general sense, it may mean metaphor, analogy, cognitive scheme, and so on. At the most concrete, it may mean a physical object, or graphical representation, or perhaps, a computer simulation. The word "consciousness" likewise has a spectrum of meanings. In Part Two, we will present our model for cosmic consciousness in the largest sense. It is a mathematical model, and this Part includes the results of a computer simulation of the model.

Our model is novel in existing outside of space and time (in some abstract realm from which space and time are created) and also in being digital (not analog, discontinuous) In Part One we locate these novel features in the context of the history of thought, and explain some of our motivations. In Chapter 1 we review the historical approaches of Western philosophy involving the concept of a continuous field of consciousness – soul, spirit, and so on. In Chapter 2 we treat likewise of the approaches of Eastern philosophy – *ākāśa*, *cakra* (chakra), *kośa* (kosha), *tattva*, and so on. In Chapter 3, we outline the parallel history of digital fields, and in Chapter 4, the quantum vacuum. Finally, in Chapter 5 we review the scientific literature on paranormal phenomena, including the various theories of continuous mental fields that have evolved in that literature.

Chapter 1.
Consciousness Models, Western

In the philosophical tradition of the West (from its roots in Ancient India, Egypt, and Mesopotamia, and further development in Greece and Europe), consciousness models involve hierarchies of continuous fields, such as the intellectual sphere, or the World Soul. In this chapter, we trace a concise chronology of the World Soul.

The individual soul is an ageless idea, attested in prehistoric times by the oral traditions of all cultures. But as far as we know, it enters history in ancient Egypt. We will begin with the individual soul in ancient Egypt, then recount the birth of the World Soul in the Pythagorean community of ancient Greece, and trace it through the Western Esoteric Tradition until its demise in Kepler's writings, along with the rise of modern science, around 1600 CE. Then we tell of the rebirth of the World Soul recently, rising from the ashes of, again, modern science.

1.1. Ancient Egypt, 2500-600 BCE

We take seriously the possibility that ancient Egyptian culture began around 10,000 BCE. Thus, tentatively, we may regard it as the Ur source for the soul concepts of the Western Esoteric Tradition, including the Greek and the Indian roots. We begin our story of the history (as opposed to the prehistory) of the soul in 2500 BCE, with the Great Pyramid of Cheops.

> This world is alive in its entirety and infused with divine spirit ... [3]

---
[3](West, 1995; p. 46)

> [Spiritual elements or bodies of ancient Egypt include the Ba, the Ka, and several others. The Ba, or soul, is] the animating principle, the vital or divine spark that vivifies all sentient creatures ... [4]
>
> [The Ka, or double, is] the power that fixes and makes individual the animating spirit that is Ba ... [5]
>
> If during life on earth, the Ka has degenerated to the point where it has been divested of all virtue, of everything truly human, then it does not reincarnate, and the Ka disperses into the various lower animal and vegetal realms. ... It may be this understanding that lies behind the curious doctrine of metempsychosis in which the deceased may be reborn as an animal or even a bush or tree.[6]

Recent studies of the Pyramid of Cheops and the pyramid texts give an idea of the journey of the soul in the reincarnation process. "After death, the Pharoah's soul was said to become a star, to join with Orion in the sky."[7] Alexander Badawy determined in 1964 that the two shafts, cut 200 feet from the King's chamber to the surface, were aimed at the Pole star, and Orion, in the year 2600 BCE. And according to Robert Baumol, the two shafts from the Queen's chamber to the surface were aimed at Orion and Sirius in 2450 BCE.[8] The supposition is that these shafts were to facilitate the journey of the Pharoah's soul to its home in the sky, after the death and internment of his body in the pyramid.

---

[4](West, 1995; p. 63)
[5](West, 1995; p. 64)
[6](West, 1995; p. 64)
[7](West, 1995; p. 452)
[8](West, 1995; p. 453)

1.2. Ancient Greece, 800 BCE – 500 CE

Ancient Greek philosophy evolved in part from Ancient Indian, Egyptian, and Mesopotamian roots.[9]

1.2.1. Homer and Hesiod, ca 800 BCE

According to legend, Homer was a blind poet, and author of the epic poems, the *Iliad* and the *Odyssey*. Hesiod, contemporary or a generation later, also wrote epic poems. They wrote the earliest surviving Greek poetry. We owe to Homer, perhaps, the earliest historical theory of soul. The *psyche* (breath-soul) and the *thymos* (blood-soul) were the two organs of consciousness. They converged, in the centuries following Homer and Hesiod, into a single soul concept.[10] The ancient Hebrews also had twin concepts, *nephesh* (soul) and *ruach* (spirit), which migrated into Early Christian theology.[11]

1.2.2. Anaximander of Miletus, 610-546 BCE

Anaximander was the author of the earliest extant Greek prose.[12] He was inspired by Mesopotamian ideas.[13] The is a cosmological theory created by him in the 6th century BCE. His work is mostly lost.[14] From the few extant fragments, we learn that he believed the beginning or first principle (*arkhē*) is an endless, unlimited mass (*apeiron*), subject to neither old age nor decay, which perpetually yields fresh materials from which everything

---

[9](McEvilley, 2002)
[10](Paul Lee, private), (Snell, 1960), (Onions, 1951; pp. 2, 23, 99, 116)
[11](Onians, 1973; p. 480)
[12](Kahn, 1985; p. 240)
[13](McEvilley, 2002; p. 30)
[14]Kahn, 1985

which we can perceive is derived. The apeiron was never precisely defined, and it has generally been understood (e.g. by Aristotle and Augustine) as a sort of primal chaos. It embraced the opposites of hot and cold, wet and dry, and directed the movement of things, by which there grew up all of the host of shapes and differences which are found in the world. The primeval chaos idea may have derived from Hesiod's *Theogony*[15] or from the Orphic trinity, *Chaos, Gaia, Eros*.[16]

1.2.3. Pythagoras of Samos, 572-497 BCE

Pythagoras of Samos was born around 570 BCE. He traveled and studied in Egypt and Babylon. Initiated into the mysteries of several traditions — Egyptian, Babylonian, and Persian — he returned to Greece and Magna Graecia in southern Italy and carried on with the reforms set in place by the Orphic religion, which became the most important religion of ancient Greece. He is reputed to have been a student of Anaximander. Pythagoras synthesized spiritual and natural philosophy into the framework for classical Greek culture, including the metaphysical and sacred aspect of Number, the One (monad, unity), and its emanations. He introduced the terms philosophy and cosmos. He created a school around 520 BCE in Croton (southern Italy) that emphasized communal living, gender equality, vegetarianism, mystery initiations, Orphic poetry, harmonics, music therapy, the monochord, geometry, arithmetic, and cosmology. The school was destroyed by a rejected and disgruntled follower who led a popular revolt against the community around 500 BCE. Among the important followers of Pythagoras were Philolaus (474 - 385 BCE) and Archytas of Tarentum (428 - 347 BCE), an important influence on Plato.

The Pythagorean doctrine is based on these three principles:

---

[15](Abraham, 1994; p. 131)
[16](Abraham, 1994; p. 93)

1. Ideas: matter is attracted to absolute forms, or ideas, which have an existence of their own. Mathematics is the study of these forms.
2. Souls: an animal has an immortal soul, which reincarnates (transmigrates) after death, until a state of perfection is attained.
3. Harmony: ideas and souls are related by sympathy, resonance, or musical ratio.

We may recognize the Pythagorean theory of reincarnation as derived from the Egyptian. The idea of the World Soul evolved in this community. The Pythagorean emphasis on the number One suggests an influence of Indian monism.[17]

1.2.4. Heraclitus of Ephesus, 535-475 BCE

Heraclitus was a pre-Socratic philosopher from the coast of present-day Turkey. He is known for the aphorism, *everything is in a state of flux*. Everything is in flux, but while flowing, maintains identity. The cosmos is derived from a single principle, the *Logos*. Similar ideas are found in Indian philosophy some centuries later. He was an important influence on Plato.

1.2.5. Parmenides of Elea, 515-450 BCE

Following the *One* of Pythagoras and the *Logos* of Heraclitus, Parmenides derived the cosmos from a single principle, Being, or *eon*, from which emerges cosmic consciousness, or *nous*.[18] Being is outside of space and time.[19] He described two kinds of reality: *the Way of Truth*, and *the Way of Opinion*. This may

---

[17](McEvilley, 2002; pp. 44-48)
[18](Geldard, 2007; p. 45)
[19](Geldard, 2007; p. 103)

be interpreted as a form of mind/body duality.[20] He was also an influence on Plato.

1.2.6. Socrates, 479-399 BCE

Socrates was the agent of a major shift in which classical philosophy turned from nature (or *physis*) to human life, and rational consciousness was firmly established. Also, he is considered among the first to emphasize the concept of the World Soul. He is known primarily from the portrayal of him in Plato's dialogues.

1.2.7. Plato, 429-347 BCE

Plato synthesized Socrates and Pythagoras. First he became a follower of Socrates. He had the genius to grasp Socrates' meaning, and to present it brilliantly in a series of ten dialogues. Around 390 BCE, Plato had visited Western Greece (Southern Italy and Sicily), encountered Pythagorean communities, met Archytas of Tarentum, the great Pythagorean, and adopted Pythagoreanism as a second influence. Platonism consists in the joining of these two streams, the Socratic and the Pythagorean. In 387 BCE, Plato created his school in Academe, a suburb of Athens.

Plato expanded the teaching of Socrates on the perfection of the soul into a complete system. In this system, morals and justice were based on absolute ideas. Wisdom consists of knowledge of these ideas, and philosophy is the search for wisdom. In fourteen more dialogues, Plato elaborated this unified system. His monistic cosmology, emanating from The One, or the Good, derived from Indian sources[21].

---
[20](Geldard, 2007; p. 100)
[21](McEvilley, 2002; Ch. 5)

Plato's theory of soul is set out primarily in six of the dialogues: Phaedo, Republic II, and Phaedrus, of the middle group of dialogues, 387-367 BCE, Timaeus, around 365 BCE, which divides the middle and last groups, and Philebus and Laws, of the last group, 365-347 BCE.

The development of the individual soul is given in the three dialogues of the middle group, Phaedo, Republic II, and Phaedrus. The Phaedo is a long and detailed examination of the individual soul, its immortality, and reincarnation, supposedly given by Socrates on the day of his death sentence. The Republic describes Plato's mathematical curriculum for the Academy: arithmetic, plane geometry, solid geometry, astronomy, and music. At the end is the Tale of Er, which details the reincarnation process of the individual soul, as told by an eye witness. In the Phaedrus, Socrates and Phaedrus discourse on love, and on rhetoric. To understand divine madness, one must learn the nature of the soul. Soul is always in motion, and is self-moving, and therefore is deathless. Then begins the important metaphor of the chariot: two winged horses and a charioteer. This metaphor of the soul is used to explain divine madness, and the dynamics of reincarnation.

The *World Soul* is developed in the three later dialogues, Timaeus, Philebus, and Laws. The Timaeus is a discussion among four persons: Socrates, Timaeus, Critias, and Hermocrates. It begins with a review by Socrates of a discussion on the preceding day. This concerned the constitution of the ideal State and its citizens. Then Critias tells the famous story of Atlantis, which was told to his great-grandfather by Solon, one of the seven sages. Then Timaeus is asked to begin the feast with a description of the creation of the Universe. He tells how God, because he was good, made the world after an eternal pattern. He brought order into the world, and soul and intelligence. The world is composed of fire and earth. Being solids, these two elements require two more, water and air, to bind them. The world is a sphere with the soul in the center. The gods made man and the lower animals, and God made the human souls of the same four elements as the body of the universe, along with part of the soul of the universe. Then he set in motion the process of incarnation and reincarnation of these human souls in mortal bodies. The created gods make these mortal bodies of the four elements. As a person becomes a rational creature through education, his human soul moves in a circle in the head (a sphere) within his mortal body.

The Philebus is a lecture by Socrates on wisdom and pleasure. Along the way, he introduces the World Soul as the source of individual souls. The Laws is the last of Plato's writings. It is a long dialogue of three older men, and is unique in that Socrates is absent. The actions of the World Soul are discussed in detail.

The relation between Platonic philosophy and India has been studied for many years by Thomas McEvilley, and occupies four chapters in his book, *The Shape of Ancient Thought: Comparative Studies in Greek and Indian Philosophies*.[22]

1.2.8. Aristotle, 384-322 BCE

Aristotle proposed a detailed theory of the soul, a sort of immaterial form for living things, in *De Anima*. This became a center of controversy for Christian theology in the middle ages.[23]

1.2.9. The Stoics, ca 300 BCE

From Plato and Aristotle and their followers came the Stoics, for 500 years the leading school of Greek philosophy. Among other ideas, they further developed the *logos* concept of Heraclitus,

---
[22] (McEvilley, 2002)
[23] (Rubenstein, 2003)

Aristotle, and Philo. This became an element in the Neoplatonic cosmology of Plotinus. Logos has many meanings. Cognate of the verb *legein*, to say, it may mean language, speech, expression, explanation, formula, purpose, rational structure, plan.

Following Aristotle, the Stoics adopted two principles, or archai: one active, the other passive. These are body and soul, or matter and logos.[24] For the Stoics, logos makes the world by giving form to matter in a dynamical process. Like Plato, the Stoics believed that the cosmos was a living being, with a World Soul.

1.2.10. Plotinus, 204-270 CE

The main stimuli for the Neoplatonism of Plotinus were Plato, the Middle Platonists, and to a lesser extent, the Stoics. From Plato came Plotinus' main cosmology of the three primal hypostases: the One, the Intelligence (or Intellectual Principle), and the World Soul. For Plotinus the logos was a supplementary structure that intertwined the three hypostases. He defined it as "a power that acts upon matter, not conscious of it, but merely acting upon it." This Neoplatonic cosmology, further developed by Porphyry, Iamblichus, Proclus, and others, may be regarded as the main trunk of the Western Esoteric Tradition.

1.2.11. Proclus, 412-485 CE

Proclus came to the Platonic Academy as a student, studied with Plutarch and Syrianus, and stayed for life. He was an outstanding mathematician as well as philosopher, and was the last of the great Athenians. His version of the Neoplatonic cosmology is rather ornate. He has, as Plotinus, the three hypostases: the One (En, the Henadic Realm), Being (nous), and the Soul.

---

[24]((Hahn, 1977; pp.29, 61, 74)

The nous is divided in three parts.[25] The World Soul (including individual souls) is placed between the Soul hypostasis and Nature (including embodied individual souls).[26] Proclus had a strong influence on Renaissance Neoplatonism.

1.3. Middle Ages, 800-1400 CE

During the Middle Ages, philosophy prospered in the intellectual milieu of Islam, especially in the Sufism movement.

1.3.1. al-Kindi, 805-873 CE

al-Kindi was an Islamic heir of Plato and the Neoplatonists. For him, the World Soul was an emanation from the One, as light from the Sun. His astrological work, *De radiis* was an important influence on the western scientists Roger Grosseteste (1168-1253), Roger Bacon (1214-1294), Marsilio Ficino (1433-1499), and John Dee (1527-1609). *De radiis* presented an astrological theory based on rays from the planets. Everything radiates, and space is full of these radiations.

1.3.2. Suhrawardi, 1153-1191 CE

Suhrawardi restored the ancient Greek philosophy of light, and early Persian angelology, within Islam. He connected Plato and Zoroaster.[27] He was greatly influenced by Proclus.

---

[25](Proclus, 1987; p. xxii)
[26](Proclus, 1970; p. xviii)
[27](Corbin, 1977; pp. 12, 110)

1.3.3. Ibn al-'Arabi, 1165-1240 CE

, representing the high point of Sufi philosophy, was much influenced by Neoplatonism, and perhaps by Hinduism as well.[28] Among his principles were the Oneness of Being (as in Anaximander and Parmenides) and the Creative Imagination (similar to the emanations of Plato).

1.3.4. Roger Bacon, 1214-1294 CE

Some of Aristotle's works were widely read in the Christian Middle Ages, others were unknown. But in the 12th century, Christian workmen in Muslim Spain discovered the missing works in Arabic. Due to conflicts with the Christian dogma of the sacraments, these were banned in Europe until the middle of the 13th century.[29] When this ban expired, Aristotle revolutionized European science, especially in the works of Roger Bacon, sometimes regarded as the first modern scientist. In *De multiplicatione specierum*, around 1267, Bacon presents a doctrine of the physics of light. He was inspired by Plotinus, Al-Kindi, and Roger Grosseteste. This doctrine survived for three centuries, and ended with John Dee. The word *species* meant the likeness of any object, transmitted through any medium.

1.4. Renaissance, 1400-1600 CE

Renaissance Neoplatonism began with a revival and expansion of the Greek idea of the World Soul. But the Renaissance ended with a rejection of the concept.

---

[28](Ibn a'-'Arabi, 1980; pp. 22-23)
[29](Rubenstein, 2003)

TABLE 1.1 Ficino's Cosmology

Collective	Individual Discarnate	Individual Embodied
The One (*to en*)		
The Intelligence (*nous*)	Reason	Ideas
The Soul (*psyche*)	Angels	Individual soul (incl. mind)
Spirit (*pneuma*)	Stars	Individual spirit
Nature (*physis*)	Matter	Body

1.4.1. Ficino, 1433-1499 CE

Ficino's originality derived from the syncretism of Pagan and Christian elements effected under the impulse of Plato, Plotinus, Proclus, the Hermetica, the Areopagite, Augustine, and Aquinas, to name only his primary wells of inspiration. Among the facets of this syncretism were:

- orphic music, music therapy (Ficino's personal practice),
- astrology (astrological psychology),
- magic, psychology.

He was heir to the long line of astrological magic — Synesius, Proclus, Macrobius, and Al Kindi — and was followed by Bruno and Agrippa. His cosmological model combined Neoplatonic and Christian elements, and set the foundation for the whole of Renaissance philosophy. It is summarized in the Table 1.1. The One is the undivided source of everything. The Intelligence, or Cosmic Mind, contains Plato's ideas, the archetypes and blueprints for creation. The Soul has three parts (rational, sensitive, and vegetative) and gives rise to individual minds, both human and angelic. Reason communicates between the Intelligence and the Soul, and Spirit (astral matter) intermediates between the World Soul and Nature, the created universe of matter, energy, and life.

Ficino's astrological magic, psychology, and medical practice were based on his understanding of Spirit, and its relation to the stars and planets. They have a contemporary revival in the work of James Hillman and Thomas Moore.[30]

1.4.2. Gilbert, 1544-1603 CE

It is to William Gilbert that we owe our concept of a continuous physical field. The case of a magnetic field, the first field of physics, was presented in his book, *On Magnets*, in 1600. He influenced his contemporaries, Kepler, founder of the universal gravitational field, and Galileo, the first modern dynamicist, among others.

1.4.3. Kepler, 1571-1630 CE

In his work on elliptical orbits of the planets (especially Mars), Johannes Kepler proposed a theory of universal gravitation, the second field of modern physics. In his explanation of noncircular motion, he actually changed the word *spirit* (as in angelic influence) to *force* (that is, mechanism) in the manuscript for his most important work, *Astronomia Nova*, of 1609. And here we may locate the death of the World Soul, concomitant with the birth of modern physics.

---

[30](Hillman, 1996), (Moore, 1992)

1.4.4. Galileo, 1564-1642 CE

As a youngster, Galileo worked with his father, Vincenzo, on experimental aesthetics. This early deviation from the received wisdom of the ancients became his normal mode of working. It was apparent in his kinematic experiments at the University of Pisa around 1589, which contradicted Aristotelean dogma and caused him to be expelled from the faculty in 1591. His experimental method became the paradigm of modern science. Later he came into conflict with the Roman Church for his support of the Copernican model, and also for his atomic theory, again contradicting Aristotelean dogma. In his writing there seems to be no Pythagorean nor Neoplatonic elements: he is totally modern.[31]

1.5. Baroque, 1600-1750 CE

Following the enormously influential works of Galileo and Kepler around 1600, modern science came into its own, bringing substantial changes to Western philosophy and all religions. Although alchemy and astrology remained popular, they gradually declined as the rational view of science ascended. The arts experienced a major shift in style from the Mannerism of the Late Renaissance, and the philosophy of Leibniz became the organizing center of intellectual life.[32]

1.5.1. René Descartes, 1596-1650 CE

Descartes developed the mechanistic philosophy of the late Renaissance into a detalied physiology and psychophysiology, based on his notion of *animal spirits*. For example, these spirits flowed

---
[31] On Kepler, Galileo, and Descartes, see (Burtt, 1927/1959).
[32] See (Serres, 1968) and (Deleuze, 1988/1993).

from the pineal gland, an organ the size of a pea in the human brain, to the muscles to perform motion according to will. While earlier theories used spirits to fill the gap between the soul and the body, Descartes economized this trajectory by locating the soul within (or on) the pineal gland. An idea, for him, was a figure traced on the pineal by spirits.[33] The mind/body problem, beginning with Descartes, became a defining problem for our time.[34]

1.5.2. Gottfried Wilhelm Leibniz, 1646-1716

While Leibniz may be best known today for his contributions to mathematics – the coinvention (with Newton) of the calculus – in his own time he was better known for his contributions to philosophy. His rational metaphysics was presented in two books in his lifetime – the *Combinatorial Art* and the *Theodicy* – and numerous articles. The *Theodicy* of 1710 included his theory of God as the creator of the best of all possible worlds. It was (and remains) a very difficult read, and in response to many requests from his fans he provided three abstracts: *Principles of Nature and Grace Based on Reason* and *The Theodicy, an Abridgment of the Argument Reduced to Syllogistic Form* in 1710, and *Monodology* in 1914. This latter text, occupying only 22 pages in (Weiner, 1951), comprises 90 numbered paragraphs. Here we find a model of cosmic consciousness that today we would describe as a complex dynamical system. The nodes, – called *monads* after Pythagoras – include souls, to which are attached bodies. All nodes are linked directly to God, and thus to each other. Each node has an internal dynamical system, the rules of which are affected by the data from all other nodes. This provides a solution to the mind/body problem posed by Descartes, and is very similar to our model described in Part Two of this book.

---

[33] See (Gaukroger, 1995; pp. 199, 272, 281, 388) and (Shea, 1991).
[34] (De Quincey, 2002/2010; p. 79)

1.6. Modern Times, 1900 CE

The individual soul has been with us at least since 2500 BCE. But we have argued that the World Soul emerged into documented literature with the Pythagoreans, around 500 BCE, and died with Galileo and Kepler, around 1600 CE, along with the birth of modern science. It has been missed.[35] The support for our common sense of the coherence of all and everything has been lacking since modern science became our theology and cosmology. Calls for a renewed foundation for the cosmos are now multiplying, as the books of Teilhard de Chardin, Rupert Sheldrake, and others, testify.[36]

1.6.1. Vladimir Vernadsky, 1863-1945

The biosphere concept was created by Lamarck in Paris in 1802, and named by Suess in Vienna in 1875. With the concentric spheres (lithosphere, aquasphere, biosphere, atmosphere) in place, it was only a matter of time before a sphere of consciousness was acknowledged by the scientific community. This was provided by the Russian geochemist Vladimir Vernadsky. He put forward the idea in his Paris lectures of 1922, which was then further developed and popularized by Eduard Le Roy and Teilhard de Chardin in 1927. Vernadsky's book *Biosfera*, written in Paris in 1926, became a classic of wholistic science. The third edition, written in Moscow in 1943, included a new section, *Some Words about the Noosphere*. The idea diffused widely through the book by Teilhard de Chardin, *The Phenomenon of Man* of 1955.

---

[35](Abraham, 2000, 2008)
[36]See also (Combs, 2009), (De Quincy, 2002/2010), (Varela, 1992), (and Varela, 1999).

1.6.2. Merleau-Ponty, 1908-1961

Since the Renaissance, there has been an enormous development of Western philosophy, yet little that we might regard as following directly in the tradition of Ficino. One exception, however, is the French phenomenologist, Maurice Merleau-Ponty. The final essay of his career (of 57 pages), *The Intertwining – The Chiasm*, has been influential on the frontiers of cognitive science. In it, he introduces a medium, a continuous field called *flesh*, between the mind and the body, and that provides a reciprocal connection for perception.

> It is this Visibility, this generality of the Sensible in itself, this anonymity innate to Myself that we have previously called flesh, and one knows there is no name in traditional philosophy to designate it.[37]

This is similar to the *spirit*, interpolated between the soul and the body by Ficino.

1.6.3. Sheldrake, b. 1942

In his first book, *A New Science of Life: The Hypothesis of Formative Causation* of 1981, Rupert Sheldrake begins with a consideration of unsolved problems of biology, in the areas of behavior, evolution, the origin of life, parapsychology, and so on. He delineates three levels of wholism: mechanism, vitalism, and organicism. We may relate these, respectively, to Nature, Spirit, and the World Soul levels of the Table 1.1 in Section 1.4 above.

Building on the twentieth century organismic ideas of Whitehead, Smuts, Waddington, and others, Sheldrake poses the ex-

---

[37] Merleau-Ponty, 1968; Part Four, p. 139

istence of non-energetic fields, called morphogenetic fields, that direct the emergence of form in complex systems of all kinds. In the contexts of physics, chemistry, biology, and the social sciences, these may be called morphic fields, mental fields, family fields, and so on. Although non-energetic, these fields may have measurable effects on energetic systems. Sheldrake describes the effect of a morphogenetic field on an energetic system metaphorically as morphic resonance. His hypothesis of formative causation proposes that these fields evolve from unknown seeds called morphogenetic germs. Then they evolve their structures from previous similar systems; the past intervenes in the present; morphogenetic fields have memory.

In terms of the premodern cosmologies described above, we may locate Sheldrake's morphogenetic fields in the World Soul, while the morphogenetic germs reside in the Intelligence. The entire paradigm is organismic.

1.7. Maps of Consciousness

Collecting all this information from Ancient Egypt to Modern Times, we have constructed a map of consciousness according to the Western Esoteric Tradition. In summary, we have a scheme of concentric spheres as it were, more-or-less as given us by Plato. Precipitating from these global realms we also have bits and pieces corresponding to individual living beings. All this extends the notion of space into higher realms, while the ordinary notion of Time prevails throughout.

The extension into higher realms, in all of these maps of consciousness, follow a common structure, the metaphor of a hierarchy of *levels of consciousness*. The theologian Paul Tillich explains this commonality as a universal tendency of mind.

> The diversity of beings has led the human mind

to seek for unity in diversity, because man can perceive the encountered manifoldness of things only with the help of uniting principles. One of the most universal principles used for this purpose is that of a hierarchical order in which every genus and species of things, and through them every individual thing, has its place. This way of discovering order in the seeming chaos of reality distinguishes grades and levels of being. Ontological qualities, such as higher degree of universality or a richer development of potentiality, determines the place which is ascribed to a level of being. The old term *hierarchy (holy order of rulers, disposed in rank of sacramental power)* is most expressive for this kind of thinking. It can be applied to earthly rulers as well as to genera and species of beings in nature, for example, the inorganic, the organic, the psychological. In this view, reality is seen as a pyramid of levels following each other in vertical direction according to their power of being and their grade of value. This imagery of rulers (*archoi*) in the term *hierarchy* gives to the higher levels a higher quality but a smaller quantity of exemplars. The top is monarchic, whether the monarch is a priest, an emperor, a god, of the God of monotheism.[38]

Here we may interpret a discrete stack of layers as a cognitive strategy for imaging a continuum of an esoteric dimension, or conversely, we might view any continuum as a diffuse view of a stack of layers. The One may be regarded as outside of space and time, while the other global realms may be regarded as continuous spaces. As our model, the digital akasha, is discrete rather than continuous, we cannot refer to this Western philosophical tradition from Anaximander to Sheldrake for inspiration.

Meanwhile, the ideas from Indian philosophy that found their

---
[38](Tilllich, 1963; pp. 12-13)

way into the Western tradition via Pythagoras, Plato, and Plotinus and were rapidly developed there, continued to evolve in India well into the Middle Ages, reaching eventually a great richness and maturity. We now turn to the Eastern tradition to excavate these models, for they are substantially richer in detail and sophistication than those of the Western Tradition. They are closer to the mathematical models of modern computer simulation, and they have inspired our work, as we shall see.

The very recent literature of the scientific study of consciousness is also turniing to the Eastern traditions for new ideas and an integral view of the cosmos.[39]

---

[39]See (Lanza, 2009), Sobel, 1984/2007), (E. Thompson, 2007), and (W. Thompson, 1981).

Chapter 2.
Consciousness Models, Eastern

We met in 2006 at the Indian Statistical Institute in Kolkata. Sisir had been working on a mathematical model, a complex dynamical system, for the quantum vacuum.[40] Meanwhile, I had been teaching a minicourse on the computer simulation of complex dynamical systems. So it was inevitable that we began to work together on a computer simulation of the quantum vacuum, resulting in our first joint publication.[41] During our joint work on the quantum vacuum, we discovered our common interest in meditation and the models or consciousness of Indian philosophy. The special focus of the Ramakrishna Mission on science and consciousness encouraged us and led to our joint work on models for consciousness.[42] This book is devoted to our work on consciousness.

While the details of our model are postponed to Part Two, we need to explain at this point that our model starts with a *dynamical cellular network*. This consists of a very large number of *nodes* (such as the neurons in an artificial neural network or a brain) that store information and are connected by links or synapses that move information from one node to another. This network is changing very rapidly, as information flows, and nodes appear and disappear. We position the network in an abstract mathematical universe, outside of space and time. Eventually, space and time are constructed from the model by an algorithmic process. This is similar to the cosmological models of Kashmiri Shaivism, as we will soon see.

At first we had only a vague idea of the relationship between our work and the Indian tradition, but we began digging into the literature in search of parallels. With the help of Professor

---

[40] (Requardt and Roy, 2001), reprinted as Appendix 1 here.
[41] (Abraham and Roy, 2007), reprinted as Appendix 3 here.
[42] (Abraham, 2006), reprinted as Appendix 2 here.

Debabrata SenSharma of the Indology Research Center of the Ramakrishna Mission Institute of Culture in Kolkata we discovered some extraordinary parallels, especially regarding the dynamic cellular network and the creation of space and time. We tell the story of our findings from Indology in this chapter.

Indian philosophy abounds with models of consciousness comprising a hierarchy of parallel planes of existence, such as the five koshas, the seven chakras, and so on.[43] There was a climax of Indian philosophy in the Advaita Shaivism of Kashmir, around 1000 CE, which includes an extensive stack of planes, the thirty-six tattvas. It was here that we found the most significant prefigurations of our ideas. Our special attraction to Kashmiri Shaivism is well illustrated by this clipping from *The Yoga of Vasishta*.

> VASISHTA replied: There does exist, O Rāma, the power or energy of the infinite consciousness, which is in motion all the time; that alone is the reality of all inevitable futuristic events, for it penetrates all the epochs in time. It is by that power that the nature of every object in the universe is ordained. That power (cit sákti) is also known as Mahāsattā (the great existence), Mahāciti (the great intelligence), Mahāśakti (the great power), Mahādṛṣṭi (the great vision), Mahākriyā (the great doer or doing), Mahādbhavā (the great becoming), Mahāspandā (the great vibration). It is this power that endows everything with its characteristic quality.[44]

Next, we may place the Trika tradition (the Advaita Shaivism of Kashmir) in the chronology of Indian philosophy, following SenSharma (2007, Ch. 1) and (Dyczkowski, 1987).[45] Then we

---

[43] See for example (Aurobindo, 1996, 1997).
[44] Venkatesananda, 1993; p. 89.
[45] See also (SenSharma, 1990, 2003, 2004).

will go on to describe the concepts of space and time found in the Trika literature.

2.1. Trika in Context

The Yoga tradition, upon a prehistoric base attested by cylinder seals from archeological sites, developed historically in parallel with the Vedic tradition. Yogic practices were collected and organized by Patanjali around 150 BCE. Tantrism, named after written scriptures known as Tantras, became popular after the Buddha, about 400 BCE. The Tantras are a class of religio-philosophical literature emphasizing spiritual practices (sadhanakriya) with mystic, esoteric, and magical elements. There are three Hindu Tantrika traditions: Vaisnava (based on worship of Vishnu), Sakta (on worship of Sakti), and Saiva (on worship of Siva). Vaisnava and Saiva are the major streams.

Before Islam (around 1300 CE) Kashmir was an important center of learning, Hindu and Buddhist, including the development of Tantrism. Kashmiri Shaivism, from 700 or so, became the leading form of Hinduism in Kashmir. Kashmiri Shaiva texts from 850 CE or so have survived. The texts basic to Vaisnava (the agamas) are mostly lost.

There were ten Sakta schools: Kali, Tara, Sodasi, Bhuvanesvari, Bhairavi, ... Kamala. Of the 64 Sakta Tantras mentioned by Sankara around 800 CE, most have been lost. There were 8 Saiva schools, of which three remain popular today: Saiva Siddhanta, Virasaiva, and Trika Saiva. The Saiva Tantras – 10 dualistic and 18 monistic-cum-dualistic – are mostly lost, but of 64 monist texts – the Bhairava Tantras – some have survived.

The Spanda school was an early development of Kashmiri Shaivism, which evolved many subcultures, including Saivasiddhanta, Bhuta and Garuda Tantras, Vamatantras, Bhairavatantras, and others, such as the Pasupatas, Kaulas (including Krama and Trika). Kula is a system of many major traditions, including the Kaulas (and thus also, Krama and Trika).

The Trika Saiva school reached an apex with the Kashmiri master, Abhinavagupta. His magnum opus, the *Tantraloka*, written around 1000 CE near Srinigar, is the main source prefiguring our model of consciousness, the digital ākāśa. The Trika of Abhinavagupta brought together the main threads of Kashmiri Shaivism, including ancient strands – Krama, earlier Trika, and other parts of Kula – and newer developments native to Kashmir – Pratyabhijna (the philosophy of recognition) and Spanda (the doctrine of vibration).[46] The doctrine of vibration is of particular relevance to our model of consciousness.[47]

2.2. Trika Creation Story

Now we may continue with the basic concepts of Trika Shaivism, based on Abhinavagupta, as we have learned from Professor Sen-Sharma. We are especially interested in these concepts: the 36 tattvas (levels of creation), spanda (vibration), ākāśa (aether), and subtle time, which emerge in the Trika creation story. We begin with this story, in a nutshell.

The Kashmiri Saivites describe the creation of the world as the self-manifestation of the fundamental consciousness outside of the universe (Supreme Lord, Śiva). This begins with its subtle form (conscious energy, Divine Śakti), which is activated by spanda (a throb). This is like a movement, yet not a movement, as space and time do not yet exist. This activation leads to a descent or involution, all in an instant, down a sequence of four spheres (andas) in 36 steps (tattvas, or categories). On the

---

[46] See Dyczkowski, 1987, Introduction and Chapter 1), also (Odier, 2005) and (Shankarananda, 2006, Ch. 2.)
[47] (Panda, 1995)

upper sphere (Saktyanda) Śakti functions as pure consciousness (cit śakti) including levels of creation in ideal form, that is, as cosmic ideas. Subject and object are one. On the next lower sphere (Mynda) Śakti functions as the subtle power of observation, Subject and Object begin to differentiate. On the next (Prakrityanda) she functions as gross material power (Prakriti), and in the base sphere (Prithvyanda) as the most gross material power.

The tattvas are as follows.[48]

The Pure Order, which exists in the realm of Mahamaya (the pure form of Divine Sakti), has the first five tattvas:

- 1. Śiva
- 2. Śakti
- 3. Sadasiva
- 4. Isvara
- 5. Sadvidya.

The Impure Order, with the remaining 31 tattvas, is characterized by the operation of Maya (which conceals and covers up) and by limited nature, discreteness, and material form. The first of the impure tattvas are:

The six Kancukas, or truncations of Divine Powers:

- 6. Maya (beginning of limitation, discreteness, and differentiation)
- 7. Kaal (limitation of time)
- 8. Vidya (limitation of knowledge)
- 9. Raga (attachment)

---

[48]Here we follow Pandey, 2006, p. 362.

- 10. Kalaa (limitation of action)
- 11. Niyati (limitation of place)

Next, the inner instrument:

- 12. Purusa
- 13. Prakrti
- 14. Buddhi (cognition)
- 15. Ahamkara (ego)
- 16. Manas (ratiocination)

The final twenty (in four groups of five):

- five sense organs (smelling, tasting, seeing, feeling, hearing)
- five instruments of action (resting, rejecting, locomotion, handling, voicing)
- five tanmatras (smell, taste, form, touch, sound)
- five mahabhutas (earth, water, light, air, ether)

2.3. Selected Concepts of Trika

There are several concepts of Kashmiri Shaivism that are of special interest in connection with our model of consciousness.

2.3.1 Nonduality

It is said that in philosophy, there are two methods of reconciling a dichotomy: the dualistic method in which both alternatives coexist in conflict, and the nondualistic or monistic method in which one subsumes the other. In the case of the mind/body dichotomy of Descartes, the nondualistic way is sometimes called monistic idealism or integral monism.

A chief feature of Kashmiri Shaivism, shared with Advaita Vedanta, is integral monism. This is in contrast with the Samkhya philosophy, which is dualistic. The Trika cosmology is nondualistic: divine consciousness (Siva) is primary. The material world is created from it, and embraced within it. Thus subject and object are one. Dualistic theories regard the creator and the created as distinct.

Although the Trika and Vedanta are both advaita (nondualistic), there are important differences. Says Dyczkowski:

> Saivism equates the absolute wholly with consciousness. Reality is pure consciousness alone (*samvid*). Consciousness and Being are synonymous. To experience the essential identity between them is to enjoy the bliss (*ananda*) of realization. The Advaita Vedantin maintains that in a primary sense reality cannot be characterized in any particular way, but affirms that secondarily we can conceive it to be 'Being-Consciousness-Bliss' (*saccidananda*). Being, understood as an absolute substance (which is not substantial in a material sense), is the model for the Advaita conception of consciousness. Monistic Saivism, on the other hand, considers consciousness to be the basic model through which we understand Being.[49]

In our model, gross space, time, and matter are created from a rapidly fluctuating dynamical cellular network by an algorithmic process of condensation. The dynamic network corresponds to divine consciousness, the matrix of all existence, outside of

---
[49]Dyczkowski, 1987; p. 43.

conventional space and time. The rapid fluctuations correspond with spanda.

2.3.2. Spanda

Spanda (pronounced 'spund') means throb or pulse in Sanskrit. In Trika philosophy, it refers to a nonmoving motion of the Divine Consciousness with which the creation of the world is initiated and maintained. The first text of the Kasmiri Saiva tradition is the *Sivasutra*, or *Aphorisms of Siva*, attributed to Vasugupta around 850 CE. A few years later, Vasugupta's disciple, Kallatabhatta, wrote the *Spandakarika* (*Stanzas on Vibration*, The Spanda school is named after this work.[50] From Jaideva Singh:

> Spanda is a very technical word of this system. Literally, it is some sort of movement or throb. But as applied to the Divine, it cannot mean movement.
>
> Abhinavagupta makes this point luminously clear in these lines:
>
> > Spandana means some sort of movement. If there is movement from the essential nature of the Divine towards another object, it is definite movement, not some sort, otherwise, movement itself would be nothing. Therefore, *Spanda* is only a throb, a heaving of spiritual rapture in the essential nature of the Divine which excludes all succession. This is the significance of the word Kiñcit in *kiñcit calanam* which is to be interpreted as *movement as it were*.
>
> Movement or motion occurs only in a spatio-temporal framework. The Supreme transcends all notions of

---
[50]See (Dyczkowski. 1987; p. 20).

space and time. Further from Abhinavagupta:

> *Spandana* means some sort of movement. The characteristic of 'somewhat' consists in the fact that even the immovable appears 'as if moving,' because though the light of consciousness does not change in the least, yet it appears to be changing *as it were*. The immovable appears as if having a variety of manifestation.
>
> *Spanda* is, therefore, spiritual dynamism without any movement in itself but serving as the *causa sine qua non* of all movements.[51]

In our model, the iteration of a dynamic cellular network corresponds to Spanda. The nodes of the network change their states with each increment of network time, and links between nodes also appear and disappear.

2.3.3. Śiva and Śakti.

Śiva and Śakti are two principle tattvas in Shaiva darshan. Śiva is considered as *prakasa* or luminating force and Śakti as *vimarsa* or vibratory force. Here,

> Prakasa transmutes into vimarsa and assumes the form of bindu [point]. Vimarsa, or Śakti transmutes into prakasa, or Śiva, as a result of which bindu is split and nada is born. Prakasa is that energy which is inherent in Siva..... Mool bindu is the root of creation. When there is vibration in bindu, nada is born and the tattvas that arise out of nada

---
[51](Singh, 1980; pp. xvi-xvii)

are formed....there is evolution of sound into light, which gives rise to form.[52]

These are the first two tattvas of the Pure Order. Although Śakti folows Śiva in the sequence of tattvas, the two are generally regarded as a linked and balanced pair. Dyczkowski explains,

> Intimately bound together as heat is with fire, or coolness with ice, Śakti – God's power, and Śiva – its possessor, are never separate. Even so, if we are to understand their relationship we must provisionally distinguish between them in the realms of manifestation.[53]

In the context of our digital model, Śiva and Śakti might be interpreted as the alternating inflow and outflow functions of the dynamic links between the nodes.

2.3.4. Māyā

Māyā means 'not that' in Sanskrit. In Advaita Vedanta, it was introduced by Shankara in the 9th century referring to a veil that hides the Divine from us, creating the illusion of duality. In Trika, it is the first tattva of the Impure Order, the first of the six kancukas. Also, all the tattvas of the Impure Order are characterized by Māyā Śakti. SenSharma explains, in the creative process (involution):

> As the second type of *cidanus* [spiritual monads] undergo involution in the domain of *Māyā śakti*,

---
[52]number 6
[53]Dyczkowski, 1987; p. 99

which is described as the universal power of obscuration, she enwraps them with the result that their natures get further obscured. The veiling by *Māyā* is technically called the *mayiya mala*.

Māyā is not alone in accomplishing the task of obscuration. It brings into operation five other forces of limitation, technically called *kancukas*. As these *kancukas* (lit. integuments) enwrap the individual being, Śiva's divine powers as the Supreme Lord, which were indicative of His divine glory, are transformed into five principles of limitation (*kancukas*) ...[54]

2.3.5. Ākāśa

In Vidya, Tattva 8 belonging to the Māyā group, Śiva creates space and time. The spatial and material universe is created of five elements: four of matter (fire, earth, air, water), and one of pure space or ether (the *ākāśa*).[55] The ether has two phases, subtle and gross. In our model, the nodes of the network correspond to the subtle ether, while the physical space created in temporal slices by condensation corresponds to the gross ether. The nodes condense into *cliques* that precipitate as *fuzzy lumps* in an illusion of continuous three-dimensional space. These lumps provide the granularity of space (around the Planck scale, or smallest spatial scale according to quantum physics) required by the theory of the quantum vacuum.

2.3.6. Time

In the same tattva, Śiva creates two times: subtle and gross. Likewise, our model has two times: the subtle time of the dy-

---

[54](SenSharma, 2007; p. 75)
[55] Also spelled akasa or akasha.

namical network, and the gross time created in the condensation process.⁵⁶ Together with the continuous-appearing granularity of space around the Planck scale, the condensation process creates a submicroscopically granular gross time with an illusion of continuity, completing the creation of continuous spacetime based on certain dynamics. In the beginning, there were no geometric forms, but they emerge from the network. This is similar to the space associated to the ākāśa tattva.

This complicated double structure of space and time seems far from our intuitive notions, perhaps too complicated. However, the recent development of quantum philosophy is moving in a similar direction, as attested from Karen Barad, one of the leading contemporary theorists on the foundations of quantum physics.

> Intra-actions are nonarbitrary, nondeterministic causal enactments through which matter-in-the-process-of-becoming is iteratively enfolded into its ongoing differential materialization. Such a dynamic is not marked by an exterior parameter called time, nor does it take place in a container called space. Rather, *iterative intra-actions are the dynamics through which temporality and spatiality are produced and iteratively reconfigured in the materialization of phenomena and the (re)making of material-discursive boundaries and their constitutive exclusions.*⁵⁷

Well yes, professional philosophy has a language as arcane as mathematics. But this quote from a contemporary philosopher of science joins medieval Kashmiri Shaivism in support of the novel features of our model. Next, we discuss the ākāśa in the larger picture of Indian philosophy.

---

⁵⁶(Balslev, 2009)
⁵⁷(Barad, 2007; pp. 179)

2.4 Ākāśa

The word *ākāśa* is generally translated in English as *ether*. However, the concept of *ether* as a medium for the propagation of light is not the appropriate one in Indian philosophy. The concept of akasa has the distinguishing quality of sound in contrast to *ether* as that of light. In fact, to understand the concept of akasa, one needs to understand the concept of *tattva*. The Sanskrit word *tattva* consists of two syllables: *tat* and *tva*. *Tat* means *that* and *tva* means *ness* and hence the word *tattva* signifies *thatness*. On further analysis, it signifies *the essence which creates the feeling of existence*.[58] There is another Sanskrit word *bhuta* which is used synonymously with *tattva*.

The emergence of the akasha occurs in the final group of five tattvas, the mahabhutas, These are tattvas numbers 32 – 36. These are known as the *pancha tattva* or *pancha mahabhuta* and are associated with distinct vibratory motions. These motions appear during the evolutionary process of manifestation from Parabrahman. The first evolutionary state is the ākāśa tattva. It has the distinguishing quality of sound. If we want to use a word similar to ether for ākāśa it is better to use an adjective with ether. The five tattva can be classified as:[59]

- Ākāśa tattva as the sonoriferous ether (sound)
- Vāyu as the tangiferous ether (touch)
- Tejas tattva as the luminiferous ether (color)
- Apas as the gustiferous ether (taste)
- Prithvi as the odoriferous ether (smell).

Evolution gives rise to light from sound and then to forms. The generation of light from sound has been discovered in twentieth

---
[58](Saraswati, 1984; p. 24).
[59](Prasad, 1989; p. 1).

century physics and the phenomenon is known as *sonoluminescence*.[60]

On the gross level, the physical characteristics of these five tattvas or five mahabhutas can be described as:[61]

- The characteristics of akasa are motion in all directions which are not agglomerated and also not obstructed.
- Tejas corresponds to fire, i.e., going upward, burning, lighting, shining, destruction, power.
- The characteristics of vayu or air are movability and friction.
- Apas corresponds to water which characterize smoothness, softness, heaviness, coolness, purification, etc.
- Prithvi or the earth corresponds to form, stability, rigidity, support, etc.

According to Samkhya philosophy, atoms of the five mahabhutas combine together to form different substances. According to the different schools of Indian philosophy, matter can exist in three forms as tanmatras (i.e. sub-atomic stage), as anus, or the atoms of the mahabhutas. The tanmatras signify the potency of having the characteristics of akasa, fire, air earth, etc. A divergence of views exists regarding the genesis of the tanmatras. Actually, they possess something more than the quantum of mass and energy, they possess the physical characteristics like penetrability, capability of radiation of heat, viscosity and cohesion. In addition to these capabilities, they also possess the potentials of energies represented by sound, touch, colour, taste, and smell, but are devoid of any particular form. In this way both animate and inanimate bodies and all forms are created out of the various combinations of these five elements or pancha mahabhutas.

---

[60]Putterman, 1995; p. 46-51).
[61]Dasgupta, 1974; p. 222).

Akasa is all pervading, just like the luminiferous ether described in physics. The vibrations of the elements which constitute sound associated with akasa are different from the vibrations which produce sound and require a physical medium. These elements or tanmatras are very subtle but have the potentiality of creating the sound in the physical world under certain conditions. These subtle tattvas exist in the universe on four planes as follows:

- Physiological, corresponding to prana
- Mental, corresponding to manas
- Psychic corresponding to vijnna
- Spiritual corresponding to ananda

Again some of the secondary qulaities of these tattva can be summarised as:

- Space: this is considered to be a quality of the akasa tattwa. The vibration here may give rise to the statistical nature of space.
- Locomotion: a quality of vayu tattva, motion in all directions.
- Expansion: a quality of tejas tattwa.
- Contraction: a quality of ap tattva. The direction of this ether is considered to be the reverse direction of the ether associated to tejas tattva or agni tattva.
- Coherence resistance: a quality of the prithvi tattva. This is opposite to akasa tattwa. Akasa tattwa can give rise to locomotion where as prithvi tattva resists it,

It is worth mentioning that Laszlo[62] proposed an integral theory of everything and the importance of the akasic field in several

---

[62]Eg, (Laszlo, 2004).

of his recent monographs, as we describe in a later chapter. We shall now briefly summarise the various schemes of tattvas according to different schools of Indian philosophy.[63].

- Kashmir Shaiva Siddhanta and Shaiva darshan: there exists thirty six tattvas of creation.
- Nyaya philosophy: sixteen tattvas.
- Samkhya philosophy: twenty-five tattvas.
- Vaishesika: six tattvas.
- Advaita Vedanta: one tattva.
- Dvaita Vedanta: two tattvas.
- Vishishta Advaita: three tattvas.

While there is a better fit between our model and Eastern philosophy than there is with the Western tradition, we have not yet discovered a justification for the atomic or quantum aspect of our model. So, we turn to that in the next chapter.

---

[63](Saraswati, 2008; p. 80)

Chapter 3. Atomism

The idea of atomism has been eleborated and extensively discussed in the philsophical traditions of East and West for many centuries. The concept of the discreteness of spacetime at Planck scale revives ancient atomism in the milieu of modern physics today. Here are some of the works from antiquity on.[64]

3.1 Ancient Greece

As far back as we can see, the atomism thread in Ancient Greece begins with the presocratics.

Parmenides, 450 BCE

According to Popper (1998), Parmenides of Elea – an important presocratic philosopher of Ancient Greece – was the creator of atomism (*atomos*, Greek for indivisible). First of all, he is known for his *Two Ways – The Way of True Knowledge* (*aletheia*) and *The Way of Human Conjecture* (*doxa*) – revealed to him by a goddess and described in his only work, *On nature*. *The Way of True Knowledge* includes the idea that behind the false and illusory world of change perceived by the senses there is an absolute reality that is totally static, a dark sphere of continuous dense matter, called Being, or *eon*.[65] In our sensory perceptions, we experience a dual world of atoms moving in the void, hence *The Way of Human Conjecture*.

---

[64](Bell, 2005)
[65]See (Geldard, 2007; p. 109).

Democritus, 400 BCE

Democritus, a student of Parmenides, is sometimes regarded as the founder of atomism. It is said that Democritus' ideas were formed to contradict Parmenides. Democritus wrote on math, astronomy, and ethics, and had a great influence on later Greek philosophy, especially Aristotle, and hence, on the whole of the Western Tradition.

Regarding atoms, he believed that material bodies were formed as temporary composites of eternal atoms, like flocks of birds. Atoms are variously shaped and sized. The primary qualities of a material body – its shape, size, and weight – and its secondary aspects – smell, taste, etc – all derive from the size and shape of its atoms. Atoms move in a "void", which is empty, and yet is not nothing. The soul is made of soul-atoms, which are very small and spherical, and can pass through solid material bodies, like neutrinos.

On chasing wild geese

A *wild goose chase* is English slang for a hopeless quest, perhaps due originally to Shakespeare. This comes to mind when we try to trace an idea back to its origin. Yet this is what historians love to do, and the *history of ideas* is wild goose chasing elevated to the level of an academic profession, since the prototypical case of Arthur Lovejoy, in his book, *The Great Chain of Being*. In fact, Part One of this book is an excavation of one of the great chains, and now, we are digging up the great chain of atomism.

The contemporary French philosopher Gilles Deleuze is a great chaser of wild geese. In his book *Différence et Repetition*, Chapter IV, Ideas and the Synthesis of Difference, he defines *Idea* (always written with a capital letter) and discusses the conditions under which an Idea may emerge.

> The application of these criteria must therefore be sought in very different domains, by means of examples chosen almost at random.
>
> *First example: atomism as a physical Idea.* Ancient atomism not only multiplied Parmenidean being, it also conceived of Ideas as multiplicities of atoms, atoms being the objective elements of thought. Thereafter it is indeed essential that atoms be related to other atoms at the heart of structures which are actualised in sensible composites.[66]

So we see that atomism, since its birth in the 5th century BCE, refers not only to the space-time-matter level of physics, but also to the higher realms of consciousness. For cosmic consciousness includes a neural network of individual minds, each seething with ideas, and each idea is a molecule of atoms of thought. We may as well regard the entire cosmos as a complex dynamical system, and we do!

3.2 Oriental and Islamic Views

It is always a pleasure to follow a thread from Ancient Greece, through trade routes to India, then circuitously to Early Islam, and thence to Europe. There is a long history of atomism in India, in both Hinduism and Buddhism.

On the other hand, the natural philosophy as propounded by the Chinese seems to be more inclined to synechism (the tendency to regard everything as continuous) rather than atomism. The battle between synechism and atomism was revived by Islamic scholars during later years. Next we shall describe the different Indian views on atomism.

---

[66] (Deleuze, 1994; p. 184)

3.2.1 Indian Philosophy

The six distinct philosophical systems which accept the authority of Veda are Nyāya, Vaiśesika, Sāmkhya, Yoga, Pūrvamimāmsa, and Uttara-mimāmsā or Vedanta. Kanāda is supposed to be the propounder of atomism within the Vaiśesika system. In fact, the word Kanad is derived from the word Kana, which means atom.

Nyāya Vaiśesika and Jaina Views on Atomism

The concepts of *Avayavin* (whole) and *Avayava* (part) are the key concepts in the atomic theory of the Nyāya Vaiśesika system. Here, the atom is supposed to be indestructible, indivisible, and without magnitude. Substances in this universe are thought to be of nine classes. Among these classes, four (earth, water, fire and water) are considered to be atomic. These classes have their own atomic class with particular attributes. For example, an atom of earth has odour, one of air has touch, etc.

The concept of atom as developed in Jaina is different from that in Nyāya Vaiśesika. The smallest object that can not be divided further was called *paramānu* in Jaina. However, it has the capacity to change under certain conditions.

Jaina philosophy advocates the atomic conception of time where these atoms are distinct and can never be combined. Since they cannot be combined, they cannot be classified as *astikaya*. The concept of astikaya plays an important role in Jaina philosophy. The word *astikaya* consists of two words; *asti* (exists) and *kaya* (body, or more precisely individual particles). These individual particles or bodies are supposed to mix up or be added to make a substance. In this sense, time is not classified as an astikaya. This helps to differentiate between atoms of space, matter, etc., and those of time. Time atoms in the Jaina framework are ultimate, absolute, and real. This is different from the conventional concept of time, as in minutes, hours, days, months, years, etc. The divisions and subdivisions in conventional time are linked to the different units of measurements which are generally based on some changes in the physical universe, and hence, never to be unconditional, whereas the time-atom represents unconditioned or absolute time. According to Jaina philosophy, absolute time is described as real existent and being potent, i.e. it brings the changes in other substances, birth, growth, decay of things, etc. Here, conventional time presupposes absolute time.[67]

Atomism in Buddhist Thought

Among Buddhist traditions, Vasubandhu and Dharmakirti (around 650 CE) particularly discussed the existence of atoms. Dharmakirti was a student of Dignaga, a Buddhist logician, and professor at the famed Nalanda University. He introduced in the thread a wondrous novelty, namely, that atoms are not eternal, but rather, flash into and out of existence as points of energy. This seemed somewhat outré until very recently, when the quantum vacuum emerged into physics, as we discuss in the next chapter.

Yogācāra, Mādhyamika, Vaibhāsika and Sautrāntika are the four major schools of Buddhism. The first two belong to the Mahāyāna sect and the last two to Hinayāna. The Hinayāna schools admit the concept of atomism where as the Mahāyāna schools deny it. According to these Buddhist schools there are two types of atoms, namely *Dravyaparamānu* (simple) and *Samghātaparamānu* (compound). They consider that eight classes of atoms compose the substances in our universe. Four are fundamental atoms associated with earth, water, fire and air. The other four as secondary atoms associated with colour, odour, taste, and touch. Here, one can think of the qualities in terms of atoms also. It is to be noted that Buddhists do not consider atoms in terms

---

[67](Balsev, 2009; p. 73)

of particles of some stuff, but rather in terms of force or energy. The concept of an atom as a dynamic force within the Buddhist framework is compatible with the doctrine of momentariness, where every thing is momentary, and all things change. Here, the universe is simply a *process* and a system of interconnected activities, and all is in ceaseless motion.[68]

3.2.3 Views from China

Around 290 BCE, the atomistic view was refuted in *The Book of Master Chuang*. However, Mohists around 330 BCE considered the idea of atoms or instants of time. This is evident from the following passages:

- The "beginning" means an instant of time.
- Time sometimes has duration and sometimes not, for the "beginning" point of time has no duration.

3.2.3 Islamic Thought

A controversy about atomism began soon after 800 CE between the Islamic philosophers Nazzam, a divisionist, and Abu l-Hudahyl al-Allaf, an atomist. The arguments of Avicenna (980-1037) against material atomism had considerable influence on the thinkers of medieval Europe. The most influencial proponant of atomism in Islamic schools was Motekallamin of the 10th and 11th centuries. The views are summarized in Maimonides (1135-1204). Regarding the existence of time atoms it is stated: *Time is composed of time-atoms, i.e. of many parts, which on account of their short duration cannot be divided.*

---
[68](Guenther, 1989)

3.3 The Baroque

Early Renaissance philosophy was dominated by Christian Neoplatonism, and relied on metaphors of continuous spaces and fields. In the Baroque epoch, atomism was revived.

Galileo, 1623

As described in Chapter 1, Galileo was condemned by the Vatican in 1633, overtly because of supporting Copernicus (that the earth moves) in his book, *Dialogues concerning the two chief world systems*, published in 1632. However, there is a competing (and controversial) theory according to which his real offense was his earlier book, *The Assayer*, of 1623.[69] This work advocated an atomic theory, according to which (rather like Democritus) the secondary qualities of matter (taste, smell, etc) were determined by the primary qualities (the shapes of atoms comprising the matter). This was of huge concern to the Vatican in that Transubstantiation – the official dogma of the Church since the Council of Trent (1545-1563) regarding the consecration in the Mass of the Sacraments (turning the bread and wine into the body and blood of Christ) – depended on secondary qualities being independent of primary qualities.

Further, the reduction of the secondary qualities to geometry was central to Galileo's program of the mathematization of nature,. This began *the crisis of European science* – according to Husserl – which was completed by Descartes.[70]

---

[69](Redondi, 1987)
[70](Husserl, 1954/1970; Part II)

3.4. Modern Quantum Theory

In this book we are presenting a model of consciousness. A mathematical model is a metaphor, a cognitive strategy. It is not a claim on the structure of reality. Our model is intended to be educational, and to present a vision of things that is beyond the usual frame of discussion. The model, as explained briefly at the beginning of Chapter 2 and in detail in Part Two of this book, is a mathematical structure known as a dynamical cellular network. It is outside of space and time, and yet physical space, time, energy, and matter are derived from it. In other words, in this view, consciousness is primary, the universe secondary. In this we concur with the Advaita Shaivism philosophy of Kashmir, and other monist systems.

The dynamical cellular network concept has evolved in the physics of the quantum vacuum. Yet we do not mean to imply that consciousness is identical with the quantum vacuum, as some have done.[71] Rather, we regard the mathematical structure of cosmic consciousness as similar to the mathematical structure of the quantum vacuum. Ideas may pop out of the cosmos into individual mental spacetime, while elementary particles pop out of the quantum vacuum into physical spacetime. Physicalists may regard the mind as a component of the brain, but we do not.

Here, we briefly review the emergence of the quantum vacuum in modern physics, and its recruitment as a model of consciousness. For the early history we rely heavily on the splendid work of Boi.[72]

---

[71] See, for example, (Laszlo, 1995), and our next chapter.
[72] Page numbers refer to (Boi, 2009). Also see Quantum Mechanics in the Wikipedia, and (Von Neumann, 1955; Ch. 1).

3.4.1 The beginnings

Shortly following the early death of Descartes, Newton's universal theory of gravitation laid atomism to rest, as continuous space and time provided the physical foundation for gravity as a force field, and in fact, for all matter. Atomism remained in the shadows for two hundred years. Then it rose from the ashes in a sequence of developments, collectively known as the quantum revolution. Here is a chronology of some of these developments.

- 1808, John Dalton posed a unique atom for each element
- 1838. Faraday, cathode rays
- 1859. Kirchhoff, black-body radiation
- 1877. Boltzmann. discrete energy levels
- 1897. Michelson-Morley experiment (p. 51)
- 1897, J. J. Thompson discovered the electron (Nobel prize in 1906)
- 1900, Max Planck proposed energy quanta, founded quantum theory
- 1905, Albert Einstein introduced the photon as a corpuscle
- 1913. Bohr, model of the atom, Nobel Prize 1922
- 1914, Max Planck, black body radiation
- 1920. Einstein, denies the ether (p. 57) Nobel Prize 1921
- 1924. de Broglie, all matter wave-like, Nobel Prize 1929
- 1924, Arnold Sommerfeld, quantum laws
- 1925. Heisenberg, matrix mechanics (w/ Born, Jordan)
- 1926. Heisenberg, uncertainty principle (at Bohr institute) Nobel Prize 1932
- 1926. The golden age of quantum mechanics (p. 57)
- 1926. Pascual Jordan, quantum field theory
- 1926, Ervin Schrodinger, wave mechanics
- 1927. Bohr and Heisenberg, Copenhagen interpretation

And now we come to the beginnings of the quantum vacuum.

- 1926. Paul Dirac, relativistic quantum mechanics
- 1927. Paul Dirac, the vacuum as a sea of zero-energy photons (p. 64) Nobel Prize 1933
- 1927, Dirac, Pauli, Weisskopf, Jordan, Quantum field theory
- 1928. Dirac develops the relativistic theory of the electron (p.64)
- 1930. Dirac proposes the vacuum as a sea of negative-energy electrons (pp. 51, 64)
- 1930. Dirac, QED (perturbation theory of the quantum vacuum) creation and annihilation of quantum particles, particle-antiparticle pairs, virtual particles, vacuum polarization publication of "Principles of Quantum Mechanics"
- 1931. Dirac, magnetic monopoles
- 1932. Dirac proposes the positron (p. 51)
- 1935. Einstein, EPR paradox
- 1939. Sidney Dankoff works on quantum electrodynamics (QED)
- 1940s, Feynman, Schwinger, Tomonaga, QED
- 1848. Shin'ichiro Tomonaga, renormalisation, QED (p. 66) Nobel Prize 1965
- 1948. Casimir effect (p. 59)
- 1949. Richard Feynman, QED (p. 66) Nobel Prize 1965
- 1951. Julian Schwinger, renormalization, QED (p. 66) Nobel Prize 1965
- 1951. Schwinger identifies space-time with the quantum vacuum (p. 58)
- 1966, H. Yukawa, Non-local Field Theory and Quantum Vacuum (QV)
- 1973. E. P. Tryon, vacuum fluctuation model of the universe, zero-point energy, (pp. 53, 55)

Following the development of quantum electrodynamics (QED) in 1973, came the QED theory of the 1930 quantum vacuum (QV) of Dirac. This is basic to the model of Requardt and Roy in 2001, describing the QV as a complex dynamical system. This view of nature has the vacuum full of activity, in which particles jump out from, and then back into, the vacuum, in pairs.

In QED, physicists would calculate the probabilities of transitions with respect to the vacuum state. The vacuum as such does not contribute to the calculations. The seat of a particle is a point. However in 1966, Yukawa proposed the concept of a non-local field, where the seat of a particle is an extended region, in contrast to QED. There is no distinction between empty and occupied seats. Effectively, Yukawa introduced a new quantum theory of the ether with globular structure.

3.4.2 Dirac and the quantum vacuum, 1930

The origins of the quantum vacuum evolve from the positron, predicted by Dirac in 1928, and first observed by Carl Anderson in 1932. We find the first description by Dirac in his textbook of 1930 (revised last in 1967), *The Principles of Quantum Mechanics*. Here is the historic development of the QV theory, in the original, technical, text of Dirac.

Clue #1. (In the 2nd edn of 1935) In Chapter 11, Relativistic Theory of the Electron, in the final section, Sec. 73, Theory of the Positron, we find:

> It has been mentioned in Sec. 67 that the wave equation for the electron admits of twice as many solutions as it ought to, half of them referring to states with negative values for the kinetic energy ... In this way we are led to infer that the negative-energy solutions of (56) refer to the motion of a new kind

of particle having the mass of an electron and the opposite charge. Such particles have been observed experimentally and are called *positrons*. We cannot, however, simply assert that the negative-energy solutions represent positrons, as this would make the dynamical relations all wrong. ... We must therefore establish the theory of the positrons on a somewhat different footing. We assume that *nearly all the negative-energy states are occupied*, with one electron in each state in accordance with the exclusion principle of Pauli. An unoccupied negative-energy state will now appear as something with a positive energy, since to make it disappear, i.e. to fill it up, we should have to add to it an electron with negative energy. We assume that *these unoccupied negative-energy states are the positrons*. ... A perfect vacuum is a region where all the states of positive energy are unoccupied and all those of negative energy are occupied.

Clue #2. In the fourth edition (signed 11 May 1957) the chapter on QED (Chapter 12) has been substantially revised. In the Preface, Dirac wrote:

> In present-day high-energy physics the creation and annihilation of charged particles is a frequent occurrence. A quantum electrodynamics which demands conservation of the number of charged particles is therefore out of touch with physical reality. So I have replaced [Chapter 12] by a quantum electrodynamics which includes creation and annihilation of electron-positron pairs.

Clue #3. In a revision of the fourth edition dated 26 May 1967, Dirac added two final sections in Chapter 12: *Section 81. Interpretation*, and *Section 82. Applications*. From Section 81:

> The ket $|Q>$ represents a state for which there are no photons, electrons, or positrons present. One would be inclined to suppose this state to be the perfect vacuum, but it cannot be, because it is not stationary. ... Let us call the state Q represented by $|Q>$ the no-particle state at a certain time. If we start with the no-particle state it does not remain the no-particle state. Particles get created where none previously existed, their energy coming from the interaction part of the Hamiltonian. ... which causes transitions in which a photon is emitted and simultaneously an electron-positron pair is created.

3.4.3 Quantum Gravity, 1980s

Beginning in the 1980s, physicists began inventing novel ways to integrate general relativity and quantum theory. To date, none of these attempts, collectively know as *quantum gravity*, has been successful. But the main candidates all support the idea of discrete space.[73]

3.4.4 Fredkin and the digital philosophy, 2000

The cellular automaton (CA) ideas of Stan Ulam and John von Neumann in the 1950s rested in obscurity until the appearance of John Conway's *Game of Life* in the 1970s. Then CA models of nature became a fad, and many successful models for macroscopic physical systems were made, especially in the circle around Feynman in the 1980s.[74]

However, computer science models of the individual soul are

---
[73](Smolin, 2001; p. 95)
[74](Hey, 1999)

rare, or almost nonexistent.[75] This may be due to the development of the history of science in the West stuck in the groove of Descartes' rational and reductionist worldview. However, we now seek a science of consciousness, so mathematical models must be created, and this is the locus of the creative work we present in this book.

In this connection we must mention the work of Ed Fredkin, one of the pioneers of the digital philosophy, and the mainstay of the website www.digitalphilosophy.org, which explains:

> Digital Philosophy (DP) is a new way of thinking about the fundamental workings of processes in nature. DP is an atomic theory carried to a logical extreme where all quantities in nature are finite and discrete. This means that, theoretically, any quantity can be represented exactly by an integer. Further, DP implies that nature harbors no infinities, infinitesimals, continuities, or locally determined random variables.

In *On the Soul* (2000 Draft Paper) Fredkin proposed a computer science definition of the soul, concluding: "The soul is an informational entity, which is constructed out of the states and the arrangements of material things." This seems to be a material theory, favored by many Western physicists, in which thoughts are regarded as quantum states of the physical brain, as described in the next chapter. But Fredkin is more radical, in that he proposes that thoughts are ultimately configurations of numbers. Our model, although discrete and numerical, is not a material theory. All these recent developments, which we subsume under the classical heading atomism, support the idea that underlying our illusion of continuous space, time, matter, energy, etc (the analog part of the analog/digital dichotomy, and the wave part of the wave/particle duality) is a fundamental

---
[75] One exception may be found in (Wolfram, 2002).

layer that is finite, discrete, and intelligent (that is, law-abiding). Sometimes all this is called *the finite nature assumption*.[76] This is close to the view of Parmenides described above.

In this chronology of modern physics, we may see a sequence of waves, including General Relativity (GR), Quantum Field Theory (QFT), Quantum Electrodynamics (QED), and the Quantum Vacuum (QV). The birth of the QV is found in Dirac's work, 1927-1932. We turn now, in the next chapter, to the adaptation of the QV as a model for consciousness, in the works of Fritjof Capra, Fred Alan Wolf, and others.

---

[76](Fredkin, 1992)

Chapter 4.
Quantum Theory and Consciousness

The connection between quantum theory and consciousness has several threads. The measurement theory of von Neumann and the quantum brain models of Hameroff and Penrose have been of historical interest, while the quantum vacuum theory of mind is of special relevance to this book. In this chapter we survey these threads.

4.1. Measurement Theory

Measurement theory has several stages of evolution: von Neumann (1932), Eugene Wigner (1961, 1963, 1976), Henry P. Stapp (1993, 2007), and ongoing.

4.1.1 John von Neumann (1903-1957)

In 1932, John von Neumann put quantum mechanics on a firm mathematical basis in his book, *Mathematische Grundlagen der Quantenmechanik*, published in English in 1955. This contained not only the mathematical formalism of unitary operators on Hilbert spaces, in Chapter II, but also a theory of measurement, in Chapters V and VI. Here he created an interpretation of quantum physics (called the orthodox interpretation) in which the conscious will or free choices of the experimentalist played a crucial role. He introduced the dual nature of quantum mechanics in two processes, called Process 1 and Process 2:

> Another type of intervention in material systems, in contrast to the discontinuous, non-causal and instantaneously acting experiments or measurements,

is given by the time dependent Schrodinger differential equation. This describes how the system changes continuously and causally in the course of time, if its total energy is known.[77]

This text is written for mathematical physicists, and we are indebted to a few critical readers for commentaries. For example, from Henry P. Stapp,

> In John von Neumann's rigorous mathematical formulation of quantum mechanics the effects of these free choices upon the physically described world are specifically called 'interventions' (von Neumann 1955, pp. 358, 418). These choices are "free' in the sense that they are not coerced, fixed, or determined by the physically described aspects of the theory. Yet these choices ... have potent effects ... [78]

And according to Menas Kafatos and Robert Nadeau,

> In 1932, John von Neumann developed another version of quantum measurement theory. In this version, the assumption is that both the quantum system and the measurement devices are describable in terms of what Bohr viewed as only one complementary aspect of the total reality – the wave function. In the absence of a mechanistic description of when and how the collapse of the wave function occurs, von Neumann concluded that it must occur in the consciousness of human beings.[79]

---

[77]von Neumann, 1955; p. 347.
[78]Stapp, 2007; p. 10.
[79](Kafatos, 2000; p. 38)

In other words, the time-continuous aspect of a quantum system is supported by a rigorous mathematical theory, the time-discontinuous measurements are not.[80] As there is as yet no theory for the *collapse of the wave function*, this is called the *measurement problem*.

4.1.2 Eugene Wigner (1902-1995)

In 1961, Eugene P. Wigner (Nobel Prize in Physics, 1963) published the first of three commentaries on the von Neumann interpretation. Here he asked the question: If consciousness plays a role in quantum theory, whose consciousness is it?[81] Then in 1963, he studied the weaknesses of the orthodox interpretation, and concluded,

> The principal conceptual weakness of the orthodox view is, in my opinion, that it merely abstractly postulates interactions ... For some observables, in fact the majority of them (such as xyp_z), nobody seriously believes that a measuring apparatus exists.[82]

Finally, in 1976, Wigner analyzes further problems of the orthodox interpretation.[83]

4.1.3 Henry P. Stapp

In *The Mindful Universe* of 2007, Henry Stapp presents a deeply insightful review of the orthodox interpretation of Von Neumann, informed by a detailed understanding of the mathematics

---

[80] For additional commentary, see (Redei, 2001) and (Stapp, 2007; App. B).
[81] See (Stapp, 2007; App. C).
[82] (Wheeler, 1983; p. 338)
[83] (Wheeler, 1983; p. 297), (Wolf, 1981; Ch. 13)

involved. This continues for some 84 pages, until Chapter 13, which presents his own interpretation. This significantly extends the work of von Neumann and Wigner, and connects them with the philosophy of Alfred North Whitehead. His interpretation connects significantly with our model of consciousness as well. In summarizing Whitehead's process ontology, Stapp says,

> The central idea in Whitehead's philosophy is his notion of process ... Thus in Whiteheadian process the world of fixed and settled facts grows via a sequence of actual occasions. The past actualities generate potentialities for the next actual occasion, which specifies a new spacetime standpoint (region) from which the potentialities created by the past actualities will be prehended (grasped) by the current occasion. This basic autogenetic process creates the new actual entity, which, upon its creation, contributes to the potentialities for the succeeding actual occasions.
>
> Nature's process assigns a separate spacetime region to each actual entity, and this process fills up, step-by-step, the spacetime region lying in the past of the advancing sequence of spacelike surfaces 'now', as indicated in Fig. 13.1. [Our Fig. 4.1.]
>
> The bottom curvy line represents the (spacelike) three-dimensional surface 'now' that separates at some stage of the process of creation the spacetime region corresponding to the fixed and settled past from the region corresponding to the potential future.[84]

Following this summary, Stapp steps forward from non-relativistic quantum theory (NRQT) to the relativistic quantum field theory (RQFT) of Tomonaga and Schwinger.

---

[84](Stapp, 2007; p. 91)

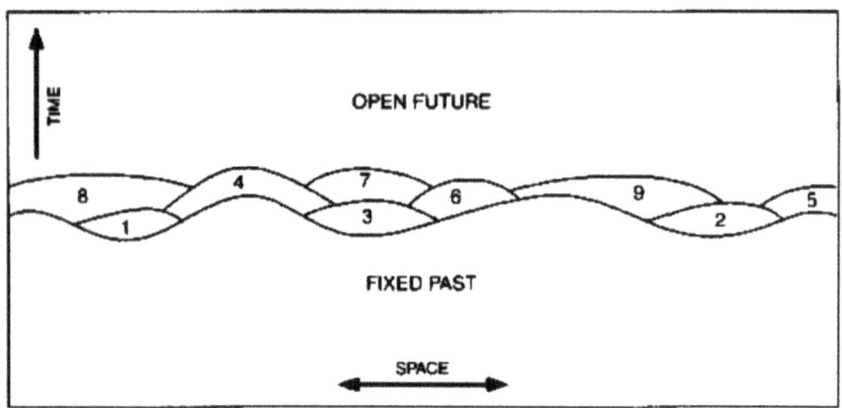

Figure 4.1. A representation of the spacetime aspects of the Whiteheadian process of creation.[86]

> In the relativistic case the wavy line in Fig. 13.1 represents some initial surface σ, an initial NOW. In the dynamical evolution of the quantum state this surface pushes continuously forward first through the spacetime region labeled 1. This unitary evolution, via the relativistic generalization of the Schroedinger equation, leaves undisturbed the aspects of the state $\Psi(\sigma)$ associated with the rest of the initial surface σ. Then a new quantum 'reduction' event occurs. ...The evolutionary process then advances the surface NOW next through region 2, then through region 3, etc.[85]

Here we have a connection with Von Neumann's process 1 and process 2. The regions of the past/future interface comprise the *quantum foam* that is fundamental to process physics. More recently, Stapp has applied his interpretation of quantum mechanics to consciousness.[87] And this is consistent as well with our model, presented in Part Two.

---

[85](Stapp, 2007; p. 92)
[87](Stapp, 2009; Chs. 13, 14)

4.1.4 Recent works

In recent works on the foundations of quantum mechanics, measurement theory is also known as *the measurement problem*. The problem is, there is as yet no satisfactory theory of measurement! In one of the most lucid texts on the subject, George Greenstein and Arthur Zajonc write,

> In this chapter we encounter a new situation. Throughout this book, we have found quantum mechanics adequate to deal with the extraordinary nature of the quantum world as revealed by modern research. But in this chapter, we will describe a topic that quantum theory finds exceedingly difficult to deal with. Indeed, many physicists believe that the theory *cannot* deal with it; and that its failure to do so points to a grave defect in the theory.
>
> The topic to which we refer is the act of making a measurement. Which measurement? Any measurement.[88]

A measurement is interpreted as a discontinuous change in the state of the system, as described by its wave function. Described conventionally as a *collapse of the wave function*, this is one of the mainstays of the *orthodox interpretation* of quantum mechanics, as described above. Continuing, Greenstein and Zajonc write,

> We conclude that the collapse of the wave function occupies an anomalous position within quantum mechanics. It is required by the fact that observations occur, but it is not predicted by quantum theory. It is an additional postulate, which must be

---

[88](Greenstein, 2006; Ch. 8, p. 215)

> made in order that quantum mechanics be consistent.[89]

Regarding this passage, science theorist Karen Barad comments,

> This additional postulate is called the "projection postulate," and it is an ad hoc addendum to the theory introduced by the mathematician John von Neumann. ...
>
> In what sense, if any, does the projection postulate account for the "collapse"? In essence, the projection postulate is nothing more than a mathematical restatement of the alleged collapse of the wave function; that is, it is a formal statement to the effect that upon measurement we get a definite value for the property measured (that is, the measuring instrument's "pointer" points in the direction of *one* out of all possible eigenvalues). Many physicists and philosophers of physics endorse the projection postulate and take it to be a well-estabilshied feature of the so-called Copenhagen interpretation.[90]

We are going to return now to Wigner's idea, that measurement involves the entanglement of the quantum system with consciousness.[91] Note that this entanglement is an aspect of the mind/body problem, the resolution of which is one of the target goals for our model of consciousness presented in Part Two.

---

[89](Greenstein, 2006; p. 221)
[90](Barad, 2007; pp. 285-286)
[91]See also (Bernstein, 1991), (Davies, 1999), and (Herbert, 1985, 1988, 1993).

4.2. Consciousness and the Quantum Brain

In the field of consciousness studies there is a long tradition of materialism, in which there is an identification of mind (or individual consciousness) with the brain. The von Neumann/Wigner interpretation of quantum mechanics is usually applied to consciousness in this materialist context, by calling upon quantum phenomena of the brain. The microscopic quantum neural network approach dates from Schroedinger (1949) and Ricciardi and Umezawa (1967).[92]

Connections between this interpretation of quantum mechanics and Eastern mysticism reached popular awareness in 1975 through the immensely popular and influential book, *The Tao of Physics*, by Fritjof Capra. A further step in this direction soon followed, in which the brain/mind is viewed as a macroscopic quantum system. This approach has been advocated by Stuart Hameroff (1987) and by Roger Penrose (1989, 1994). A noteworthy hybrid view has been put forward by Amit Goswami in 1990. All of these recent developments leave the quantum measurement problem and the mind/body problem unsolved.

> In the past few years it has become increasingly clear to me that the only view of the brain-mind that is complete and consistent in its explanatory power is this: The brain-mind is an interactive system with both classical and quantum components. These components interact within a basic idealist framework in which consciousness is primary.[93]

This is getting close to our own view, manifest in the dynamical cellular network model for consciousness described in Part Two

---

[92]See also (Braden, 1996, 2007), (Davies, 1999; p. 32), (Jitatmananda, 2006), (Searle, 1997), (Seifer, 2008), (Walker, 2000), and (Wolf, 1981; Ch. 14).
[93](Goswami, 1993; p. 164)

of this book. However, the materialist view must assume that a mental state or thought is a physical state in the material brain. Only then may physical forces, macro or microscopic, be called upon to explain the mind/brain connection.

However, we postulate (on the global scale) a primary consciousness, the Śiva tattva, which is quantum-like, and a classical part, the ākāśa tattva, which embraces the universe of matter and energy. No physical force may mitigate between the immaterial mind and the material brain. Only mathematical connections will suffice.

4.3. Consciousness and the Quantum Vacuum

Our model is inspired by the quantum vacuum, so let us now review the precursor literature.[94] The first connection between consciousness and the quantum vacuum that we know was made by Ervin Laszlo. In *l'Ipotesi del Campo* Ψ of 1987, a continuous field (as in Chapter 3 above) is posed as the carrier of extrasensory perception. But in *The Creative Cosmos* of 1993, Laszlo made the explicit identification of this field as the quantum vacuum.[95] This idea is further evolved in eight sequel books to date,

- *The Interconnected Universe*, 1995,
- *The Whispering Pond*, 1996,
- *The Connectivity Hypothesis*, 2003,
- *Science and the Akashic Field*, 2004,
- *Science and the Reenchantment of the Cosmos*, 2006,
- *Quantum Shift in the Global Brain*, 2008,

---

[94]For a contemporary review of the physics of the quantum vacuum, see (Aitchison, 2009).

[95]See (Laszlo, 1993; p. 89).

- *Cosmos: A Co-Creators Guide to the Whole World* (with Jude Currivan), 2008, and
- *The Akashic Experience: Science and the Cosmic Memory Field*, 2009.

The background of Laszlo's innovation of 1993 is given in his Introduction to *The Creative Cosmos*,

> This study tackles what in my view is *the* fundamental question of scientific inquiry into the nature of reality. ... That question is how the things that are, have *become* what they are. ...
>
> As it happened, the insight on which this book is based came to me one quiet evening in the summer of 1986 as I was sitting in the company of a few close friends and colleagues under a sky of infinite depth and clarity on the shores of the Mediterranean. We were in a reflective mood, just recovering from the shock of the news of the passing away of a mutual friend whom we had all admired for his insight and creativity as well as for his deep humanism. As each of us recounted episodes from his rich and adventurous life, someone remarked how tragic it was that all his accumulated experience and wisdom should now have vanished without a trace. I replied, with a conviction that surprised me as much as the others, that the experience and wisdom of our friend had not vanished from this world: its trace still existed the same as the trace of all things that ever took place in the universe.
>
> We fell silent. The truth of this assertion, bold as it was, had hit us all.[96]

Further, in 2004, he wrote in an autobiographical essay,

---

[96](Laszlo, 1993; pp. 15-16)

Notwithstanding these activities and commitments, I remained faithful to my basic quest. When in 1984 I left the UN for the Tuscan Hills, I took stock of how far I got. I found that I needed to go further. ...With this in mind I set about reviewing the latest findings in quantum physics, evolutionary biology, cosmology, and consciousness research. Before long, I became convinced that Whitehead's concept of internal relations is entirely sound. The natural systems of this world are indeed strongly — internally, intrinsically and even nonlocally — connected and correlated with each other.

Internal relations also apply to the human world. They apply even to our consciousness. This was brought home to me by a personal experience that I recounted in 1993 in the Preface to *The Creative Cosmos* and will not repeat here. Although a mystical experience does not provide proof of internal relations between one's mind and the minds of others, it does provide an incentive to study the possibility that such relations exist. This consideration became part of my explorations in the years that followed. The books I produced in this Tuscan period include *The Creative Cosmos* (1993), *The Interconnected Universe* (1995), *The Whispering Pond* (1997-98), and *The Connectivity Hypothesis* (2003). In these books I marshal evidence that systems in the world are intrinsically interconnected, and suggest the reason for it. The field theory I developed supplies that reason: it argues that the connections and correlations that come to light in the physical and the life sciences, the same as the transpersonal ties that emerge in experimental parapsychology and consciousness research, have one and the same root: the subtle but entirely fundamental information field associated with the quantum vacuum, the deepest and most fundamental level of physical reality in the universe.[97]

The quantum vacuum is a seething froth sparkling with elementary particles emerging from nowhere in pairs, and after a very short time, vanishing again as they came. It is not really a continuous geometrical space, nor a field like the four fundamental fields of mathematical physics. However, associated with this seething froth is a continuous field of energy, called the *zero-point energy* (or ZPE) field. It is this field that Laszlo proposes as the physical substrate of psi phenomena, the 'entirely fundamental information field associated with the quantum vacuum.'

However the entanglement of mind and body is achieved — by the mathematical fiction of a wave function, or that of the quantum vacuum or ZPE field, that of a dynamical cellular network beyond space and time, or even by the intention of the Śiva tattva — it must be counted a sort of psychokinesis or, a paranormal phenomenon. The mind – or individual soul or monad if you wish, along with its intentions and free will – and the body – with its physiology, brain, hormones, muscles, and the like – are not connected by any presently known physical field or mechanism. We may say that they are entangled, yet that idea is not manifest in quantum phenomena such as the quantum entanglement of elementary particles. Nor is it supported by any of the usual mathematical models from the physical sciences. Therefore, we may call it a paranormal phenomenon, similar to telepathy and clairvoyance. In fact, the mind/body connection is the most familiar and least disputed of all paranormal phenomena.

So we turn now to a survey of so-called paranormal phenomena, to place the mind/body problem in its proper context.

---

[97](Laszlo, 2004b; p. 7)

Chapter 5. Paranormal Research

Coming as we do from mathematics, science, and spiritual practice, we have an interest in the acceptance of scientific results on the so-called paranormal phenomena by the scientific community, and hence, the compatibility of these results with our models.[98] In this chapter, we review the best attested phenomena, and later on, after describing our model in Part Two, we will return to the question of compatibility.

To our knowledge, there are rather few experimental scientists active today in the domain of paranormal phenomena; we will call them parapsychologists. Among the leaders who have written recent books are Dean Radin of California and Rupert Sheldrake of London. We now briefly summarize their writings.

5.1. Dean Radin

Dean Radin is a scientist with a masters degree in electrical engineering and a PhD in psychology from the University of Illinois, Champaign-Urbana. After a career in telecommunications, he shifted into consciousness research and is currently Senior Scientist at the Institute of Noetic Sciences (IONS). He is the author of two books, *The Conscious Universe: The Scientific Truth of Psychic Phenomena* (1997), and *Entangled Minds: Extrasensory Experiences in a Quantum Reality* (2006).

Radin passes over popular phenomena such as ghosts, poltergeists, contacts with the deceased, channeling entities, and so on, in favor of phenomena that are well supported by careful scientific experiments. He has pioneered the use of "meta-analysis" (in widespread use since 1985), meaning the statistical combination

---

[98]For general background, see (Broughton, 1991), (Mayer, 2007), (McTaggert, 2008), (Tart, 2009), and (Tiller, 2007).

of many experiments on the same phenomena to compute the overall probability of the phenomena, expressed in chances per million, etc. Radin, although a devout parapsychologist, has given careful consideration to the skeptics and their arguments, and presents a fair and balanced summary of results with expert use of statistical methods. He finds the most robust results, and statistical significant according to meta-analyses, in six categories of phenomena, each the subject of a chapter of his 1997 book. These are telepathy, perception at a distance, perception through time, mind-matter interaction, mental interactions with living organisms, and field consciousness. Of these six categories of phenomena, the first five are called *the big five* by the eminent parapsychologist, Charles Tart (2009). We now summarize the big five from Radin (1997).

Chapter 5, Telepathy comprises three experimental arrangements: dream telepathy, ganzfeld telepathy, and autoganzfeld telepathy. In dream telepathy experiments, a sender tries to mentally send images to a receiver who is asleep in a shielded sleep laboratory. The original series of trials were conducted by Montague Ullman and Stanley Krippner at the Maimonides Medical Center in Brooklyn, New York, from 1966 to 1972, and have been replicated numerous times by others. A total of 450 sessions were reported in the literature. In ganzfeld telepathy, the receiver is prepared in a receptive state in a sensory deprivation situation. The original trials, in the mid-1970s, were conducted by Charles Honorton, William Braud, and Adrian Parker. Twenty-eight studies were analyzed in 1985. Autoganzfeld telepathy continued the ganzfeld program, but with various improvements of design, until 1989.

Chapter 6, Perception at a Distance includes three types of clairvoyance: ESP card guessing, remote viewing, and precognitive remote perception (PRP). ESP card experiments begun in 1889 were popularized by J. B. Rhine in a book of 1934. The subject guesses which card has been hidden nearby. By now, an enormous number of experiments have testified to the effect: 3.6

milliion trials involving 4,600 subjects. Remote viewing, begun in 1882, became popular in the 1970s. A program was run by Harold Puthoff, Russsell Targ, and Edwin May, at the Stanford Research Institute (SRI) from 1970 to 1994. A sender went to a remote location which the receiver attempted to capture in a drawing. Sometimes, only the latitude and longitude of the target were given to the receiver. PRP refers to a remote viewing program conducted at Princeton University from 1978. The adjective "precognitive" was prepended to the name "remote perception" (synonymous with "remote viewing") because the targets were usually selected *after* the drawing!

Chapter 7, Perception through Time covers two kinds of precognition: forced-choice and unconscious precognition. Forced-choice tests ask the subject to guess which of a fixed number of targets (eg, ESP cards) will be selected later. Charles Honorton and Diane Ferrari analyzed all forced-choice experiments published in English from 1935 to 1987, including two million trials by 50,000 subjects. Unconscious precognition refers to the detection of a future event by an unconscious nervous system response. In the presentiment experiments conducted recently by Dean Radin himself, the subject sits in a comfortable chair, her fingers connected to apparatus that records her skin conductance, heart rate, and blood volume. When she is ready, the computer selects an image from a fixed set of 120 photos, waits five seconds, then shows it to her on a screen for three seconds. There is a ten second rest before the next cycle begins. The analysis of the data sometimes shows a reaction to an arousing image, in the five seconds before the image appears on the screen. This is called *presentiment*. As in the PRP experiments reported in Radin's preceding chapter, presentiment is an occurrence of information from the future appearing in the present, violating the usual scientific belief in causality.

Chapter 8, Mind-Matter Interaction reviews tests of mental influence on inanimate matter: dice, and physical random number generators (RNGs). Tests of people trying to influence the fall of dice have been ongoing since 1935. Radin and Diane Ferrari analyzed the results of these trials up through 1989, 2.6 million throws of the dice by 2,569 people. Tests of people trying to mentally influence an RNG (eg, radioactive decay times) were pioneered by Helmut Schmidt in the 1960s at the Boeing Laboratories. Many more studies were done at Princeton University. In 1987, Radin and Princeton professor Roger Nelson analyzed the data from nearly 600 studies.

Chapter 9, Mental Interactions with Living Organisms considers mental influence on animate matter: remote healing, telepathic excitement, and the feeling of being stared at. A study of the massive literature on remote healing is presently ongoing. Extensive studies of mental influence of human physiology have been pioneered by William Braud and Marilyn Schlitz from 1974 to 1991. The sender was instructed at random times to send to the receiver, whose skin conductance was being recorded, an arousing thought, a calming thought, or no thought. The feeling of being stared at has been the target of studies since 1913, including the new studies by Rupert Sheldrake discussed below.

After describing all this history and analysis in great detail, Radin (1997) reports the results of meta-analyses of each group of experiments. One measure of the strength of the effect under study is in terms of the "statistical significance", p. For example, $p = 0.001$ means that there is only one chance in 1000 that the found results could be obtained by chance. In other words, if the result concerns 100 throws of a die, then the 100 throws should be repeated 1000 times to obtain the result by chance. Radin usually expresses p in the form, for example, "Honorton's autoganzfeld results overall produced odds against chance of forty-five thousand to one." This means that $1/p = 45,000$, or $p = 1/45000 = .000022\ldots$, or about 2.22 times ten to the power (-5). We will use the symbol e to indicate the negative exponent of statistical significance; in this case it is 5.

Results as odds against chance from Radin (2006) are shown in

Table 5.1. Here we summarize Radin's results simply by giving the value of e for the meta-analysis of each effect. These are listed in order of strength, with the most significant effect first.[99]

TABLE 5.1. e values for the strongest effects.

Effect	Studies	Trials	e
Dice PK	269	26 million	76
Conscious staring	65	34,097	46
Ganzfeld psi	88	3,145	19
Dream psi	47	1,270	10
RNG PK	595	1.1 billion	3
Unconscious intention	40	1,055	3
Unconscious staring	15	379	2
Combined	1,019	1.126 billion	104

All these effects are synchronous, that is, they do not involve precognition. We might also mention one asynchronous effect, presentiment, which has $e = 5$.[100]

5.2. Rupert Sheldrake

Next we examine two books of Rupert Sheldrake that report on original research projects: *Dogs That Know When Their Owners Are Coming Home and Other Unexplained Powers of Animals* (1999), and *The Sense of Being Stared At and Other Aspects of the Extended Mind* (2003). *Dogs that Know* reports on many experiments. The experiment giving the title to the book is the subject of Chapter 2, on telepathy between dogs and humans.

---
[99](Radin, 2006; p. 276)
[100](Radin, 2006; p. 168)

The most extensive experiment involved a terrier, Jaytee, and his owner, Pam. The results are analyzed in Appendix B.[101] The overall statistical significance reported is $e = 6$, a very strong result for a single experiment. Like the presentiment results of Dean Radin, Sheldrakes's analysis of JayTee's time at the window shows short-term precognition.

The Sense of Being Stared At, likewise, deals primarily with one effect. The results have recently been improved to $e = 6$ by conducting trials over the internet.[102] This is consistent with the meta-analysis in Table 5.1. Sheldrake also reports telepathy trials over the internet with $e = 12$.[103] This new research strategy is likely to increase the significance of all of the paranormal phenomena.

5.3. Theory

Experimentalists, of course, like to speculate on paranormal scientific theory, especially if they have a strong scientific background. And this is the case with both Radin and Sheldrake. Sheldrake, in his first book, *A New Science of Life* of 1981, offered a theory of continuous fields called (depending on the context) morphic fields, mental fields, family fields, and so on. (He avoids using the term *psi field*.) His treatment is similar to later authors writing on the psi field, but is rather more detailed and specific than most.[104] This concept may involve a new force, presently unknown to physics. Telepathy, for example, would be understood as a resonance phenomenon between vibrations in the mental field. Like us, Sheldrake was greatly influenced by classical Indian cosmologies.

---

[101] See also (Sheldrake and Smart, 2000).
[102] (Sheldrake, 2008)
[103] (Sheldrake, 2003, 2005)
[104] See, for example, (Laszlo, 1987a).

Radin, in his first book, *The Conscious Universe* of 1997, devoted a chapter to a review of theories proposed by other scientists, with emphasis on quantum nonlocality and entanglement. In conclusion, he wrote,

> An adequate theory of psi, however, will almost certainly not be quantum theory as it is presently understood. ...Living systems may require an altogether new theory.[105]

In his second book, *Entangled Minds* of 2006, Radin again devotes a chapter to psi theories. In this case, he presents a rather complete summary in seven categories:

1. Skeptical theories
2. Signal-transfer theories
3. Goal-oriented theories
4. Field theories
5. Collective-mind theories
6. Multidimensional theories
7. Quantum-mechnical theories

Of the thirty-five pages of this chapter, five pages are devoted to categories one through six (Sheldrake's morphogenetic fields are mentioned under category 4), ten pages to category seven, which includes five quantum theories of psi (including Bohm's implicate order and the measurement theory of Von Neumann, Wigner, and Stapp), and fourteen pages are devoted to entanglement. Radin is clearly betting on entanglement, while acknowledging that the jury is still out as far as theory is concerned.

As quantum theory is incomplete in some sense, as the measurement problem is still unresolved, one may resort to metaphorical

---
[105](Radin, 1997; p. 287)

thinking, and propose a quantum-like field, something like the quantum vacuum, as the entanglement matrix, or connecting linkage, for microscopic and macroscopic phenomena. At the end of his current book, *The Akashic Experience: Science and the Cosmic Memory Field*, Laszlo writes regarding ESP,

> We begin by noting that the information that reaches the mind in an extra- or non-sensory mode does not appear to have conventional limits in space and time. Such information could have come from anywhere, and could have originated at any time in the past. This suggests that the information is not local but universal. It is distributed information in a field that is present throughout nature.
>
> This is a new and perhaps surprising hypothesis, but it's borne out by cutting-edge physics and cosmology. A universal information- and memory-field could exist in nature, associated with the fundamental element of physical reality physicists call the unified field. ... Honoring an ancient insight, this is the aspect or dimension of the unified field that I have called the Akashic field.[106]

Our own cosmological model featuring both implicate order and entanglement, albeit without using quantum mechanics nor assuming a unified field, is presented in Part Two of this book, to which we now turn. Our intention is to contribute a theory, more precisely a mathematical model, in which most paranormal phenomena – especially telepathy, precognition, and psychokinesis – may be understood, including quantum entanglement and the measurement problem.

---

[106](Laszlo, 2009; pp. 247–278)

PART TWO: Our Models

In Part One we have outlined the history of the mind/body problem from ancient times to the present, in both the Eastern and Western traditions. We gave special attention to the Indian tradition, culminating in Kashmiri Shaivism, due to its mature treatment of the mind/body problem based on an experimental science of contemplation. We also devoted much discussion to quantum mechanics, and its failed attempts to resolve the mind/body problem in the context of modern science.

In Part One we also had occasion to refer to two novel aspects of our model: it exists outside of space and time, and it is digital (discrete, or atomic). We also prepared the way for the way for these novel features by describing in some detail the history and concepts of the quantum vacuum, or QV.

In this part we present our own model of cosmic consciousness in full detail. Our model does not require the QV. However, we have derived our model from a mathematical model for the QV that is characterized by these two novel features: the model of Requardt and Roy. (We will refer to this model as the *RR model*). We have repurposed the RR model as an abstract scheme for consciousness.

So we begin the exposition of our model, in Chapter 6, with an outline of the RR model for the QV.[107] Then we go on to describe a derived model for the QV, the *AR model*, in Chapter 7. This model is accompanied by monochrome graphics (color graphics and animations are available online). Part of the motivation of the AR model is to simplify the RR model to its esssentials, and to make it clearly understandable.

In the final Chapter 8 we adapt the AR model for the QV as a model for consciousness, and apply it to the mind/body prob-

---

[107](Abraham and Roy, 2007), reprinted as Appendix 3 of this book.

lem. It is here that we bring together the historical material from Part One with the new methods of quantum physics.

In the Conclusion we will step back and assess the implications of our model. Finally, some relevant articles have been reprinted as appendices at the end of this book.

Chapter 6. The RR model

The full RR model is described in (Requardt and Roy, 2001), reprinted as Appendix 1 of this book. We will need only a brief summary of its essential features, which are collected in this chapter.

6.1. History of the RR Model

Recent developments in quantum physics (quantum gravity and string theory) have raised questions about the basic concepts of spacetime and causality at the smallest (Planck) scale. The *length and time at Planck scale* are the smallest length and smallest time increments below which no measurement is possible. The concepts of space, time, and causality lose their meaning below this scale. Spacetime behaves discretely at the Planck scale.

The RR model was created around the year 2000. Requard had been working on quantum gravity for many years and had published several papers on the discrete structure of spacetime at the Planck scale. He introduced the idea of *pregeometry* in the following sense. Discrete spatial points transition from a structure of disorder to one of order at the Planck scale in a process somewhat like phase transition in magnetic material, in which the orientations of the magnetic elements change. This kind of phase transition may happen in the case of pregeometric points.

On the other hand one of us (Roy) had been working on probabilistic geometry as proposed by Menger to understand the small-scale structure of space-time. He wrote Requardt about this approach and the thought emerged that both approaches could be combined. This became the RR model.

6.2. Some Optional Graph Theory

Graph theory is a branch of mathematics. In this theory, a *graph* is a collection of *nodes* (abstract points) connected by *links* (line segments) like a tinker-toy. A *directed graph* is a collection of nodes connected by *bonds* (links with a chosen direction indicated by an arrow-head). A directed graph may be turned into a graph by erasing the arrow-heads. A *subgraph* of graph is a selection of some of its nodes and some of its links. A subgraph is *fully connected* if every pair of nodes is connected by a link. A subgraph is *maximally fully connected* if it is fully connected, but whenever another node from its parent graph is adjoined along with all its links, the enlarged subgraph is no longer fully connected. A *clique* of a graph is defined as a maximal fully connected subgraph.

6.3. Dynamical Cellular Networks

A *dynamical cellular network* is a system similar to a neural network consisting of:

- a directed graph,
- a natural number (positive or zero integer), the *node-state*, attached to each node,
- a label, $-1, 0$, or $+1$, called the *bond-state*, attached to each bond,
- a counter, which ticks off increments of "network time", and
- rules according to which all node-states and bond-states change with each tick.

We conceptualize such a system as a flow diagram. Each node is envisioned as holding a quantized amount of information or 'charge' that is changing, step-by step, with a ticking clock that is part of the model, and keeps track of (discrete) microtime. With each tick of the clock, quanta of information will flow through each bond, like tennis balls through a pipe. Specifically, if a bond connects a node A to a node B and has bond-state $+1$, then, at the tick, information will move from A to B. The quantity of information at A will decrease, while that at B will increase, by discrete amounts that are specified by the rules of the model. Similarly, if the bond-state is -1, then information will flow from B to A. And in case the bond-state is 0, no information will flow. Finally, with each tick of the network clock, not only information will flow according to rules, but when the flows are finished, all of the bond-states will change according to another set of rules. The two sets of rules are the primary data of the model, and are spelled out in detail in the next chapter.

Here is an example of the kind of rule that might be encountered. For any two nodes of the network, say Susan and George, if there is a bond (directed link) from Susan to George with bond-state $+1$, or if there is a bond from George to Susan with bond-state -1, and also Susan's node-state (wealth) is greater than George's, then Susan will give all her wealth to George.

At any moment in the history of the network, a graph may be constructed by erasing all bonds carrying bond-state zero, and replacing all remaining bonds by links. Thus the network is shadowed by a sequence of graphs, which we call simply *the graphs of the network*.

In complex dynamical systems theory there are many different kinds of models which are similar to dynamical cellular networks. For example, there are *cellular automata*, which are a very narrow class of models introduced by von Neumann and Ulam around 1950. These have a lattice of identical discrete dynamical systems, each with finite number of states, and connections to nearest neighbors only. Then there are *spin networks*, introduced by Penrose in 1971. These are directed graphs with three links at each node, and a weight (a positive integer) on each link. Finally we may mention *graph dynamical systems*, which emerged around 2000 to model biological networks and epidemics in social networks. These have a graph of nodes with finite states, along with dynamical rules for updating the states with each tick of the clock, depending on the states of nearby nodes. A graph dynamical system is thus a generalized cellular automata, and a dynamical cellular network is a generalized graph dynamical system.

6.4. Back to the RR Model

The RR model is a two-level system, comprising two dynamical cellular networks. The model describes how macroscopic spacetime or its underlying mesoscopic substratum emerges from a more fundamental concept, a fluctuating cellular network around the Planck scale. Geometry emerges from a purely relational picture a la Leibniz. The discrete structure at the Planck scale consists of elementary nodes which interact or exchange information with each other via bonds that play the role of irreducible elementary interactions.

The microscopic level, QX is a dynamical cellular network of *nodes* and *bonds*. The macroscopic level, ST, that self-organizes from QX, is another cellular network of nodes called *supernodes* and bonds called *superbonds*. The supernodes of ST of are cliques of the underlying graphs of QX, The system of RR ends with a metric space, that is, a geometric space endowed with a ruler for measuring distances between points.

Here we will briefly describe the condensation process, by which the ST (spacetime) network is derived from the QX (quantum vacuum) network.

6.5. The Condensation Process

This process creates the ST universe from the submicroscopic QX network, which is fluctuating in its own microtime scale, outside of ordinary space and time. With each microtime tick of the network clock, the QX network is updated. We now imagine that after a rather large and perhaps variable number of these microtime ticks, the state of the universe is to be updated, or recreated, to a new state that we will call an *occasion*. The process of creating an occasion from the activity of the QX network we call *condensation*, following the early philosophers. The ongoing condensation process and its sequence of occasions create space, macrotime, and the spacetime history of objects moving through spacetime, that we experience as human consciousness.

We may never know the details of the condensation process. But we will now attempt an abstract outline. Our goal now is to envision a process with the digital QX network as input, and the analog universe as we know it as output: the condensation process.

First of all, we may assume that although only one occasion at a time is allowed in the universe, all prior occasions have been memorized. These data may be needed for the algorithmic process of condensation.

Our experience of computer graphic creation of simulacra such as science fiction films and animated videogames provides some guidance. Objects, once created, may be placed (or as we say, instanced) in spacetime in sequential occasions merely by giving the spacetime coordinates of a central point, and updated such attributes as relative size, color, texture, temperature, biochemical concentrations, or what-have-you.

It would be very convenient if we could condense spacetime for once and for all, and then use it repeatedly for successive occasions. However, no matter whether we follow the paradigm of general relativity, that of process physics, or some other upstart coming down the line in the future, we will have to deal with the algorithmic evolution of the spacetime geometry. So, the emanation of the spatial substrate must be repeated in each condensation, and must be ongoing as we speak. We have modeled this construction in a two-step process. This is the important innovation of the RR model.

First, we pose a protospace, a discrete (digital) space with characteristics of macroscopic three-dimensional space. It is another dynamical cellular network, the ST network. Like the QX network from which it is algorithmically derived, the ST network changes with every tick of the microtime clock. The algorithm is easily understood in the context of graph theory. The nodes of the ST network are cliques, groups of nodes of the QX network.

In the second step of the RR model, a continuous geometry is derived from the ST network by a smoothing process.[108] In this process, the ST nodes are regarded as *fuzzy lumps* of space, replacing the usual notion of points of Euclidean space. These fuzzy lumps were proposed in RR as the active sites of the quantum vacuum, at which elementary particles appear and disappear in particle/antiparticle pairs.

This concludes our summary of hte RR model. In the following chapter we, will present a modification of this process that we call the *AR model*. In this process, the ST network is endowed with a sort of geometry, or pseudogeometry, in which two fuzzy lumps are considered close to each other if they have a substantial overlap when seen in the ST context. RElatively more overlap translates as closer together in space.

We then embed the ST network into Euclidean three-space as isometrically as possible. That is, we try to position the fuzzy lumps as points in Euclidean space so that their Euclidean dis-

---

[108]See (Requardt and Roy, 2001), which is reproduced as Appendix 1 here.

tance approximates their pseudogeometrical distance (or fuzzy overlap) as closely as possible, using a neural network sort of approximation procedure.

6.6. Afterword on Process Physics

It seems that the search for unity in modern physics has encountered difficulties, and some experts suspect it is at a dead-end.[109] One radical alternative to the current approach is process physics.[110] Only time will tell which path will lead to victory, and a *theory of everything*. However, we cannot avoid pointing to some similarities between our model and process physics. These include: a model outside of space and time, two kinds of time (both discrete), and an ultimately reality that is somewhat like a neural network. According Reginald Cahill, a leading exponent,

> In process physics and process philosophy reality is a succession of distinct temporal states, actual occasions, where perceivable and/or detectable aspects of reality are those long-lived states that persist because they are protected from immediate decay and dissipation by their fractal topological structure, and their 'laws' of global time evolution are emergent and not imposed. Because we have an intrinsic non-local pattern recognition system, with innovation enabled by the noise implicit in the limit to self-referencing, we see that reality is analogous to the operation of neural networks, that it is mind-like.[111]

It is now time to encounter our QX network and the AR process in detail, with graphics.

---

[109](Smolin, 2006), (Woit, 2006)
[110](Cahill, 2006, 2008)
[111]See (Cahill, 2008; p. 123) and (Stapp, 2007; Ch. 13).

Chapter 7. The AR Model

Our model for the quantum vacuum, the AR model, departs only slightly from the RR model. It is simplified version of the RR model. We now summarize this model, and also illustrate it with computer graphics. This chapter is adapted from our first joint paper.[112] Further technical details are given in Appendix 3.

As in the RR model of the preceding chapter we will have macroscopic spacetime, ST, emerging from a more fundamental concept, a dynamical cellular network, QX, which is outside of space and time. The system of RR ends with a metric space, but we follow a different method to advance to a macroscopic cellular network embedded in ordinary, flat, three-dimensional Euclidean space. We would like to obtain an *isometric embedding*, that is, a mapping of our macroscopic cellular network into Euclidean space that preserves the distances between nodes, but that is mathematically impossible in general. Even though an isometric embedding is not possible, we will try to approximate one using neural network technology.

Agent-based modeling is a new style of computer programming, suitable for modeling dynamical networks, cellular membranes, and complex dynamical systems in general. There are several programming environments for agent-based modeling, and we have used one of these, NetLogo,[113] to create a computer simulation of a simplified version of the RR model, that we call the AR model. This simulation helps to understand the action of the model, and we include in this chapter some monochrome graphics created by the simulations. The NetLogo models are made available on our website where anyone may run them as applets, and we encourage this as a supplement to this text.[114]

---

[112] Abraham and Roy, 2006
[113] http://ccl.northwestern.edu/netlogo/
[114] http://www.ralph-abraham.org/articles/Blurbs/blurb119.shtml

7.1. An Outline of the AR Process

So, we now describe a NetLogo model of spacetime that self-organizes from a submicroscopic cellular network. Here is an outline of the five-step process, along with the math concepts required.

1. We begin with a dynamical cellular network, QX, with its cellular automaton-like dynamics, as described in the RR model.

2. Recall that QX consists of nodes connected by bonds (directed links, or arrows). If we drop the arrow heads and the bonds with bond-state zero, we have a graph. As in the RR model, the process leading from QX to QT at a given microtime proceeds from the graph G of the dynamical cellular network QX.

3. In our AR model we interpolate an extra step. A *permutation* is a reordering of the ordered set $(1, 2, 3, ...n)$ for some positive integer, n.. For example, the sequence (or ordered tuple) $(1, 3, 2, 4)$ is a permutation of the sequence $(1, 2, 3, 4)$. Permutations are much studied in a branch of mathematics called *combinatorics*, and are very useful in graph theory as well as other branches of math. Any permutation may be represented as a graph. Simply place the index sequence $(1, \ldots, n)$ around a circle, in clockwise order, starting at the top. Given any two of these indices, say A and B, with A preceding B in the original order, and draw an undirected link from A to B only if they are in reverse order in the permutation. This is called a permutation graph. Each permutation has a unique permutation graph, and vice versa. And now, back to our AR process.

   Rather than defining the emergent supernodes directly as the cliques of the graph G of QX, we derive from G its permutation graph, which is made as follows. Count the nodes of G. That is, list the nodes in any arbitrary order,

Assign the number 1 to the first one in this ordering, 2 to the second one, and so on. Now indicate the nodes in order around a circle, and fill in the nondirected links from the data of the graph, G.

4. We now define the supernodes of the emergent ST as the cliques of the permutation graph of P, rather than those of G. The purpose of this extension is to achieve a manageable computational task. While the computation of the cliques of a general graph is very difficult, it is relatively easy to compute the cliques of a permutation graph.

5. Spatial geometry is going to evolve from the dynamics of the QX network. For the emergence of spatial organization we use a neural network approach, based on the differences of finite sets, rather than the random metric of RR based on fuzzy sets.

And now for some details.

7.2. The QX Model

We consider a set of nodes. The number of nodes in a serious simulation would be astronomical, but for the same of illustration we will take a small number, such as 6. In general, we let N denote the number of nodes. Also, we assume that the nodes given an arbitrary ordering, $(A_1, A_2, \ldots, A_N)$, for convenience in describing (and programming) the model. The subscripts, $1, 2, \ldots, N$, etc called *indices*.

Notations

Nodes have node-states, which are interpreted as quantities of information. That is, each node has an attribute, its node-state, which is an integral multiple of a small positive number, the *quantum of information*. Choose an index number, say i, and consider the i-th node, A_i. Then we will let s_i denote the node-state (information storage) of A_i.

Now choose another index number, say k, with $i < k$, and consider the two nodes, A_i and A_k. Then A_i precedes A_k in the given ordering, and we have a bond (that is, a directed link) directed from A_i to A_k. We will denote this bond by b_{ik}. And each bond has a bond-state, J_{ik}, which is +1, 0, or −1. The bond-state may be interpreted as outgoing, off, or incoming, respectively.

Let s_{ik} denote the difference between the information s_i stored at the node A_i, and the information s_k stored at the node A_k. That is, $s_{ik} = s_k - s_i$. Sometimes we call these *node-diffs*.

In this approach, the bonds are information pipelines, and the bond-states are switches which can be switched on, off, or reversed. The wiring diagram, the pure geometry of the network, is an emergent, dynamical property and is not given in advance. Consequently, the nodes and bonds are not arranged in any regular way, and have no fixed near/far relations.

We will also have occasion to refer to the *node-weight* of a node. This is defined as the number of bonds connecting our node, A_i, to any other node, for which the bond-state is not zero. It will be denoted by w_i.

Local dynamical law

The internal node and bond states are to be updated, in discrete steps of clock microtime, according to a set of rules. The rules are the same as those of the RR model, briefly mentioned in Section 6.2 above. While various local dynamical rules might be contemplated, we are going to use just one set of rules, which is given in Definition 2.1 of the RR model.[115] Here we will paraphrase Definition 2.1. Assume two critical parameters given, $0 \leq \lambda_1 \leq \lambda_2$. Then these are the rules.

- Each node-state (information store) is increased by the net amount of incoming information from all its bond neighbors
- Each bond-state, J_{ik},
    - is unchanged if the node-state at A_i is equal to that at A_k ($s_{ik} = 0$)
    - becomes $+1$ if the difference is positive but not too much so
    $(0 < s_{ik} < \lambda_1)$
    - becomes -1 if the difference is negative but not too much so
    $(-\lambda_1 < s_{ik} < 0)$
    - becomes 0 if the difference of node state at A_i and that at A_k is too large
    $(s_{ik} > \lambda_2$ or $s_{ik} < -\lambda_2)$
    - becomes $+1$ if J_{ik} is not 0 and the difference is medium positive
    $(\lambda_1 < s_{ik} < \lambda_2)$
    - becomes -1 if J_{ik} is not 0 and the difference is medium negative
    $(-\lambda_2 < s_{ik} < -\lambda_1)$
    - becomes 0 if $J_{ik} = 0$ and the difference is medium positive or negative

Of course, we must have some initial conditions, $s_i(0)$ and $J_{ik}(0)$ in order to begin a dynamical trajectory of the cellular network.

---
[115]See Appendix 1 for the precise specifications.

Graphical displays

Our model will begin with random values for the node-states and bond-states, and then evolve with discrete steps of clock microtime according to the rules above. The node-states, s_i, node-weights, w_i, and bond-states, J_{ik}, are changing with each tick of the clock. Our computer simulations will display these data, for every tick of the clock, in a set of rapidly changing graphical displays.

Our first display will show the instantaneous bond-states of QX, J_{ik}, which take on only one of the three values, $+1, 0. -1$. Note that there are no bonds $b_{ik}(t)$, having $i = k$, as they would connect the node to itself. Therefore there are no bond-states, J_{ii} to display. Also, we only have bonds b_{ik} for $i < k$, so we only need to display the bond-states, J_{ik}, for $i < k,$. Hence the values $J_{ik}(t)$ we need to display comprise what mathematics calls an upper semi-diagonal matrix of size $N \times N$. Within this triangular display, we will indicate the three bond-state values with the color code: green for $+1$, red for -1, and yellow for 0.

We use the diagonal of the triangular matrix to show the node-states with colors: red, orange, yellow, or green, for decreasing values of node-state, s_i, which is the current charge on the $i-$th node. Alternatively, we may show the node-weights on the diagonal. All this is shown in Figure 7.1, which is a screen shot of a NetLogo simulation.

A second display shows the node-diffs in a convenient color code, above the diagonal, and the node-weights on the diagonal.

A third display shows the digraph as follows. For any $(i,k), i \neq k$, the corresponding position in the display is illuminated if there is a directed link from the $i-th$ node to the $k-th$.

The fourth and final display is the simple undirected graph underling the digraph, shown as a symmetric matrix.

7.3. The ST Model

The process by which the ST network self-organizes from QX, as described in RR, uses, as supernodes, the cliques of the graph G that underlies the dynamical cellular network. As described above, we are going to modify the prescription of RR by the addition of an intermediate step, the permutation graph, P of G.

The supernodes

Recall that the *node-weight*, w_i, of the i-th node, is the number of its adjacent nodes, t hat is, the number of bonds attached to it. Next, we form, for the $i-th$ node, the pair (i, w_i), and collect all of these in a sequence of pairs, A. Now we sort this sequence of pairs in order of decreasing weights, obtaining a new sequence of pairs, B. Finally, from B, we extract the sequence of first members, obtaining the N-permutation, P. This is a reordering of the ordered set, $(1, \ldots, N)$. We may now easily compute the cliques of the permutation graph of P as the supernodes for the ST network.

One may object that the cliques of the graph of P are not intuitively motivated, but we feel that they are at least as meaningful as the cliques of G. In fact, if we were to try to identify the cliques of G by hand, we would probably start with the nodes of highest weight.

Our NetLogo model includes a button "show permutation" that prints out, when pressed at time t, the permutation, $P(t)$. It is to export this to an external program, such as *Combinatorica*, to compute its cliques, and then to submit these to a further NetLogo model (or self-organizing map software) to obtain the ST model.

The clique computation

The cliques of a permutation graph are just the inverse sequences of its permutation, which may be found by inspection, or by software such as *Combinatorica*. We explain by considering a few examples. Here we will follow (Pemmeraju, 2003; pp. 69-71) closely, except that we use parentheses rather than brackets for vectors, that is, sequences of natural numbers.

Example 1

Let π be the permutation $(6, 5, 4, 3, 2, 1)$ of the natural sequence $(1, 2, 3, 4, 5, 6)$. Then the inversion vector of π is the 5-vector $v = (5, 4, 3, 2, 1)$. The permutation graph of π, G_π, consists of the six nodes with a link from i to j only if they are inverted, that is, $i < j$ while $\pi(i) > \pi(j)$. In this case, all nodes of G_π are linked: $6 * 5/2 = 15$ links.

In (Pemmeraju, 2003), the clique of a graph is a subset of vertices which are totally connected. We say a clique is *maximal-size* if no node may be adjoined without destroying the clique property of total connection. In (Requardt, 2001), a *clique* is always maximal-size, and we shall use this convention throughout. So in this example, there is just one clique: the entire graph is totally connected. The unique clique is the set, $\{1, 2, 3, 4, 5, 6\}$. This is a set of nodes (indices) of G_π, not of values of the permutation, π.

Example 2

Let π be the permutation $(3, 2, 1, 6, 5, 4)$. Then the permutation graph, G_π, has six links, for the inversions: $(1, 2)$ as $\pi(1) = 3 > \pi(2) = 2$, and similarly $(2, 3)$, $(1, 3)$, $(4, 5)$, $(5, 6)$, and $(4, 6)$.

There are two cliques, each of the same size, 3, which are disjoint. The permutation graph is the disjoint union of the two cliques, $\{1,2,3\}$ and $\{4,5,6\}$.

Note that the cliques of G_π correspond to maximal decreasing sequences of π, and these are observable in reading π from left to right. It is easiest to reverse the sequence of π, and read its maximal increasing sequences. In this case,
$$\mathbf{Reverse}(\pi) = (4,5,6,1,2,3)$$
from which we read immediately the two cliques, $\{4,5,6\}$ and $\{1,2,3\}$.

Example 3

Let π be the permutation $(3,6,2,5,1,4)$. In this case,
$$\mathbf{Reverse}(\pi) = (4,1,5,2,6,3)$$
from which we read immediately the two cliques, $\{4,5,6\}$ and $\{1,2,3\}$ as before.

Example 4

Let π be the permutation $(4,1,2,3,6,5)$. In this case,
$$\mathbf{Reverse}(\pi) = (5,6,3,2,1,4)$$
from which we read immediately the four cliques, $(5,6)$, $(3,4)$, $(2,4)$, $(1,4)$.

The superbonds and weights

Given a permutation arising from our simulation of the QX cellular network, we are going to define its cliques as our supernodes, that is, the nodes of our *ST* digraph. So we now need to connect these clique supernodes with bonds, the *superbonds* of our scheme. It is here that we diverge from RR, and follow a new path to precise sets and weights of entanglement, rather than fuzzy sets and random metric distances. We will use Example 4 above to illustrate the concepts.

Given a finite set of natural numbers, S, define its *span* by the interval of natural numbers,

$$span(S) = [min(S), max(S)],$$

and its *length* as the natural number,

$$length(S) = card(span(S)) = max(S) - min(S) + 1.$$

Note that the empty set has length zero.

Next, given two finite sets of natural numbers, S and T, define their *lap* by the set,

$$lap(S,T) = span(S) \cap span(T),$$

and their *lapsize* by the natural number,

$$lapsize(S,T) = card(lap(S,T)),$$

that is, the cardinality of their lap. Note that if the two sets are disjoint, then their lapsize is zero.

Similarly, we define their *span* by the set,

$$span(S,T) = span(S \cup T),$$

and their *spansize* by the natural number,

$$spansize(S,T) = card(span(S,T)).$$

Finally, we define the *weight of entanglement* of the pair (S,T) (not both empty) by the ratio,

$$weight(S,T) = 1 - lapsize(S,T)/spansize(S,T).$$

Note that the weight of two sets with disjoint spans is one. Also, if $span(S) = span(T)$, then $weight(S,T) = 0$.

We may wish at this point to modify the definition of weight in the case of two sets with disjoint spans, so that the weight may be greater than one, and actually measure the distance between the two spans.

Now let's compute the weights of pairs of the cliques of Example 4 above. Let $K_1 = (5,6)$, $K_2 = (3,4)$, $K_3 = (2,4)$, and $K_4 = (1,4)$. We will compute the symmetric matrix $W = [w_{ij} = weight(K_i, K_j)]$. Note that all the diagonal elements are zero.

We begin with w_{12}. But this is one as K_1 and K_2 are disjoint. Similarly with w_{13} and w_{14}, so we have only three weights to compute from the definitions. Here we go:

$$w_{23} = weight(K_2, K_3) = 1 - lapsize(K_2, K_3)/spansize(K_2, K_3),$$

$$lap(K_2, K_3) = span(K_2) \cap span(K_3) = \{3,4\} \cap \{2,3,4\} = \{3,4\}$$

$$lapsize(K_2, K_3) = card(lap(K_2, K_3)) = card(\{3,4\}) = 2$$

$$spansize(K_2, K_3) = card(span(K_2 \cup K_3)) = card(\{2,3,4\}) = 3$$

so finally,
$$w_{23} = 1 - 2/3 = 1/3.$$

Similarly, we compute w_{24},

$$lap(K_2, K_4) = span(K_2) \cap span(K_4) = \{3,4\} \cap \{1,2,3,4\} = \{3,4\}$$

$$lapsize(K_2, K_4) = card(lap(K_2, K_4)) = card(\{3,4\}) = 2$$

$$spansize(K_2, K_4) = card(span(K_2 \cup K_4)) = card(\{1,2,3,4\}) = 4$$

so finally,
$$w_{24} = 1 - 2/4 = 1/2.$$

Finally, we compute w_{34},

$$lap(K_3, K_4) = span(K_3) \cap span(K_4)$$

$$= \{2,3,4\} \cap \{1,2,3,4\} = \{2,3,4\}$$
$$lapsize(K_3, K_4) = card(lap(K_3, K_4)) = card(\{3,4\}) = 3$$
$$spansize(K_3, K_4) = card(span(K_3 \cup K_4))$$
$$= card(\{1,2,3,4\}) = 4$$

so finally,
$$w_{34} = 1 - 3/4 = 1/4.$$

Displaying all our weights in matrix form, we have,

$$\begin{bmatrix} 0 & 1 & 1 & 1 \\ 1 & 0 & 1/3 & 1/2 \\ 1 & 1/3 & 0 & 1/4 \\ 1 & 1/2 & 1/4 & 0 \end{bmatrix}$$

7.4. The Spatial Organization

The above simulations are preliminary to the emergence of spatial organisation. In the RR framework, the emergence of spatial organization has been formulated as a random metric space.[116] Instead, we will seek an isometric embedding of our cliques and their entanglement weights. We now have our cliques and weights, but notice that the triangle inequalities are not satisfied. These are required for geometry, demanding that the distance from a point C to a point D plus the distance from D to another point E is not less than the distance directly from C to E.

The isometric embedding problem

Even were the distances to satisfy the triangle inequalities, an isometric embedding into a Euclidean space of a given dimension

---
[116] See (Roy, 1998) and (de Gesu, 2002).

might not be possible. For example, consider the pyramid or tetrahedron, the simplest of the Platonic solids. This is a system of four nodes with all six weights equal. We may isometrically embed in Euclidean 3-space, but not in the plane. In our case, we may have a cellular system with millions of nodes and we wish to embed as isometrically as possible in 3-space, so we must adjust a random embedding by a dynamical process.

So we propose to regard the nodes and weights as a neural network, and try to embed the nodes in Euclidean space (of dimension three) such that the distances at least approximate the weights as well as possible. One technique for this process is the neural network method of self-organizing maps.[117] A simpler method, easily implemented in NetLogo, is a multidimensional variant of least squares.[118] Let us begin with a random map of the nodes into Euclidean space. Then, sum up the squares of the differences between the internodal distances and the weights. We then move the node positions in 3-space so as to minimize this sum of squares.

The method of least squares (optional)

We will illustrate this simpler method for the special case described in detail in the preceding section. This case has four nodes. As above, let $w_{12} = w_{13} = w_{14} = 1$, $w_{23} = 1/3$, $w_{24} = 1/2$, and $w_{34} = 1/4$. We are going to try to embed these four nodes in the Euclidean plane, as isometrically as possible. We begin with an arbitrary map of the nodes into the plane, assuming only that all the positions are distinct.

Let $p_i = (x_i, y_i)$ denote the current position of node K_i in the Cartesian plane, $i = 1, 2, 3, 4$, and d_{ij} the Euclidean distance between K_i and K_j. Then there is a contribution $e_{ij} = (d_{ij} - $

---
[117]See (Hagen, 1996).
[118]See (Gerald, 1970).

$w_{ij})^2$ to the square error we wish to minimize. Let E denote the total error, that is, the sum of the six pair errors, e_{ij}, for pairs $ij = 12, 13, 14, 23, 24, 34$. We regard E as a function of the eight variables, $(x_1, y_1, ..., x_4, y_4)$. We will adjust the positions so as to minimize this function, that is, to find the most nearly isometric positions. In fact, we will integrate the gradient of E by the Euler algorithm.[119]

So we must now compute symbolically the partial derivatives of E with respect to each of the eight coordinate variables. Note that E is the sum of six square terms. For any one of the eight coordinate variables, there are three of the six square terms that yield zero. For example, the square term involving p_1 and p_2, $e_{12} = (d_{12} - w_{12})^2$, has nonzero partial derivatives only with respect to the four variables, x_1, y_1, x_2, y_2.

The partial of e_{12} with respect to x_1 is

$$\partial_{x_1} e_{12} = \partial_{x_1}(d_{12} - w_{12})^2 = 2(d_{12} - w_{12})\partial_{x_1} d_{12}$$

while

$$\partial_{x_1} d_{12} = \partial_{x_1}[(x_1 - x_2)^2 + (y_1 - y_2)^2]^{1/2} = (x_1 - x_2)/d_{12}$$

and thus

$$\partial_{x_1} e_{12} = 2(d_{12} - w_{12})(x_1 - x_2)/d_{12} = 2(1 - w_{12}/d_{12})(x_1 - x_2)$$

as $d_{12} \neq 0$. Note that if $d_{12} = w_{12}$, which is the result we would like, then $\partial_{x_1} e_{12} = 0$. Likewise, if $x_1 = x_2$.

All of the partial differentiations of E with respect to the eight coordinates are very similar to this first case, we must only take care with the signs.

Thus we find the eight new coordinates, $(X_1, ..., Y_4)$, by the Euler algorithm applied to the negradient of the error, E, as follows.

---

[119] Again, see (Gerald, 1970).

For the first of the eight coordinates of the adjusted configuration,
$$X_1 = x_1 - (\partial_{x_1} E)\Delta t$$
where Δt is chosen suitably small. Using the above template for all three nonzero terms,
$$\partial_{x_1} E = \partial_{x_1}(e_{12} + e_{13} + e_{14})$$
we have,
$$X_1 = x_1 - 2\{+(1 - w_{12}/d_{12})(x_1 - x_2) + (1 - w_{13}/d_{13})(x_1 - x_3)$$
$$+ (1 - w_{14}/d_{14})(x_1 - x_4)\}\Delta t$$
The other seven adjusted coordinates are found similarly,
$$Y_1 = y_1 - 2\{+(1 - w_{12}/d_{12})(y_1 - y_2) + (1 - w_{13}/d_{13})(y_1 - y_3)$$
$$+ (1 - w_{14}/d_{14})(y_1 - y_4)\}\Delta t$$
$$X_2 = x_2 - 2\{-(1 - w_{12}/d_{12})(x_1 - x_2) + (1 - w_{23}/d_{23})(x_2 - x_3)$$
$$+ (1 - w_{24}/d_{24})(x_2 - x_4)\}\Delta t$$
$$Y_2 = y_2 - 2\{-(1 - w_{12}/d_{12})(y_1 - y_2) + (1 - w_{23}/d_{23})(y_2 - y_3)$$
$$+ (1 - w_{24}/d_{24})(y_2 - y_4)\}\Delta t$$
$$X_3 = x_3 - 2\{-(1 - w_{13}/d_{13})(x_1 - x_3) - (1 - w_{23}/d_{23})(x_2 - x_3)$$
$$+ (1 - w_{34}/d_{34})(x_3 - x_4)\}\Delta t$$
$$Y_3 = y_3 - 2\{-(1 - w_{13}/d_{13})(y_1 - y_3) - (1 - w_{23}/d_{23})(y_2 - y_3)$$
$$+ (1 - w_{34}/d_{34})(y_3 - y_4)\}\Delta t$$
$$X_4 = x_4 - 2\{-(1 - w_{14}/d_{14})(x_1 - x_4) - (1 - w_{24}/d_{24})(x_2 - x_4)$$
$$- (1 - w_{34}/d_{34})(x_3 - x_4)\}\Delta t$$
$$Y_4 = y_4 - 2\{-(1 - w_{14}/d_{14})(y_1 - y_4) - (1 - w_{24}/d_{24})(y_2 - y_4)$$
$$- (1 - w_{34}/d_{34})(y_3 - y_4)\}\Delta t$$
Notice the pattern of signs: $+++, -++, --+, ---$.

7.5. Possible Implications

The validity of the postulates of geometry has been questioned around or below Planck scale during the development of modern physics in the late twentieth century. It is worth mentioning that Riemann in 1854 discussed similar issues in connection with the validity of metrical relations in indefinitely small regions.[120] Here, we have started with a working hypothesis that a type of cellular network exists at the ultimate level of the universe from which the usual spacetime emerges. On the other hand, the people working on non-commutative geometry started with the proposition that space is pointless and a kind of non-commutativity of algebra exists at the ultimate level.[121] However, they also discussed the concept of fuzzy space at Planck scale. We have shown the emergence of spatial organization using agent-based simulations.

---

[120](Riemann, 1959)
[121](Madore, 1999)

Figure 1: The NetLogo graphics window showing bond-states and node-weights. Note that the indices (i,j) are coordinates in the display, with the first index, i, increasing horizontally from left to right, and the second index, j (recall $i < j$), increasing vertically from bottom to top. The diagonal elements, (i,i), run from the lower left to the upper right corners. The colors (which may be seen in the online computer simulations) appear here as shades of gray.

Figure 2: The NetLogo graphics window showing node-diffs and node-weights.

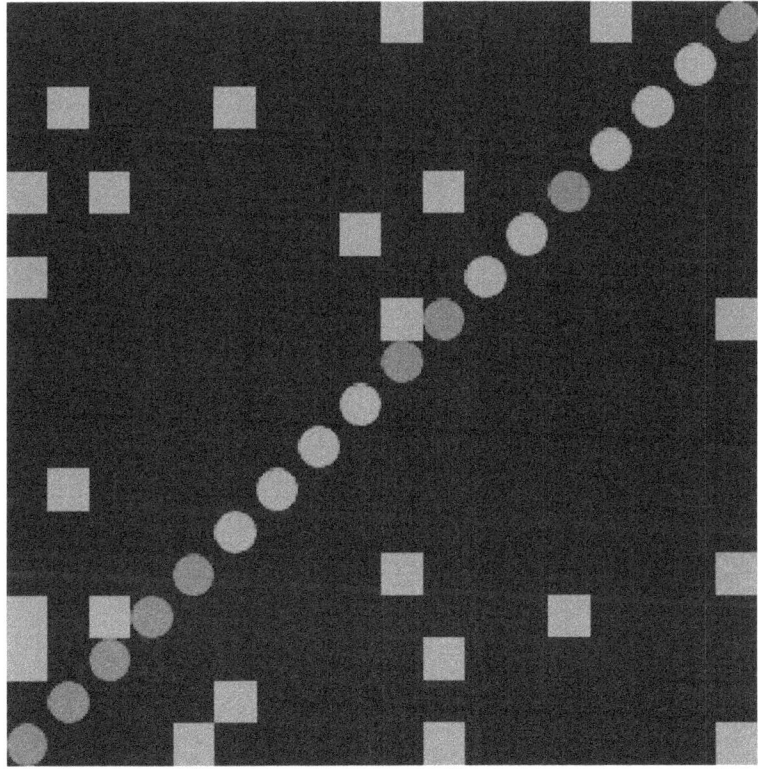

Figure 3: The NetLogo graphics window showing the digraph and node-weights.

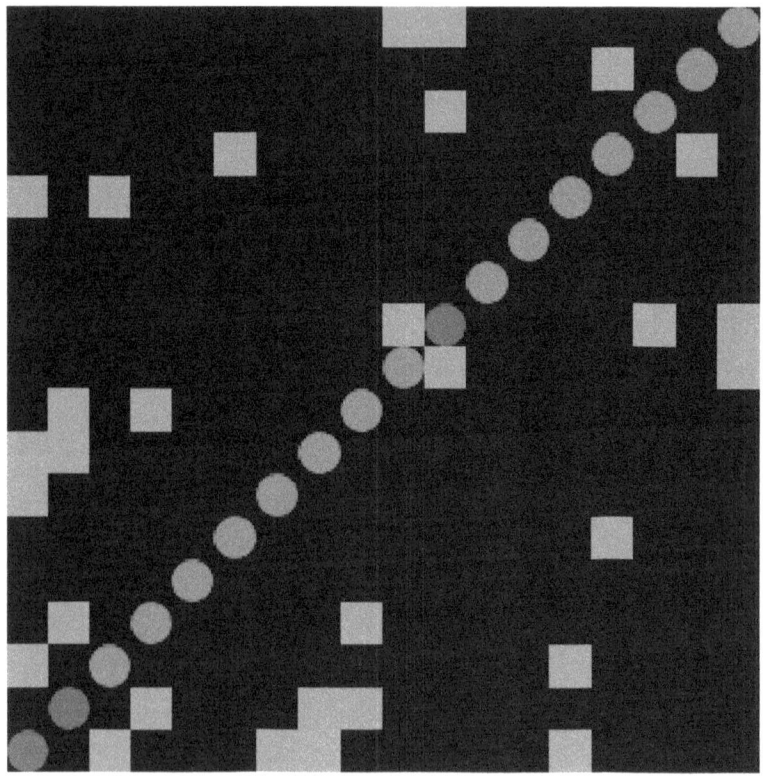

Figure 4: The NetLogo graphics window showing the graph and node-weights.

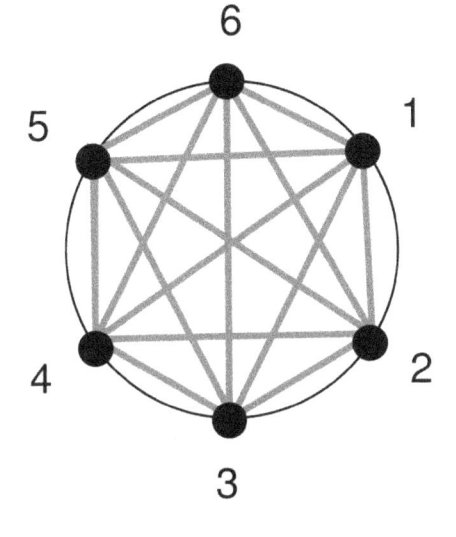

Figure 5: Permutation graph for Example 1 (one clique).

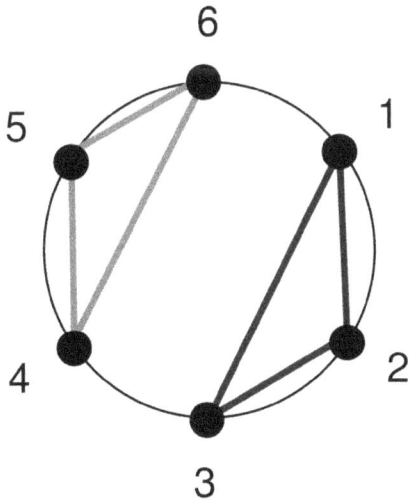

Figure 6: Permutation graph for Example 2 (two cliques).

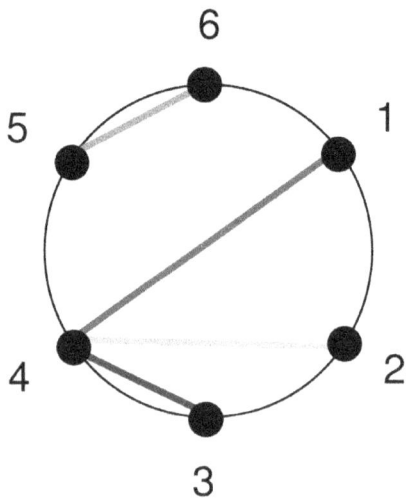

Figure 7: Permutation graph for Example 4 (four cliques).

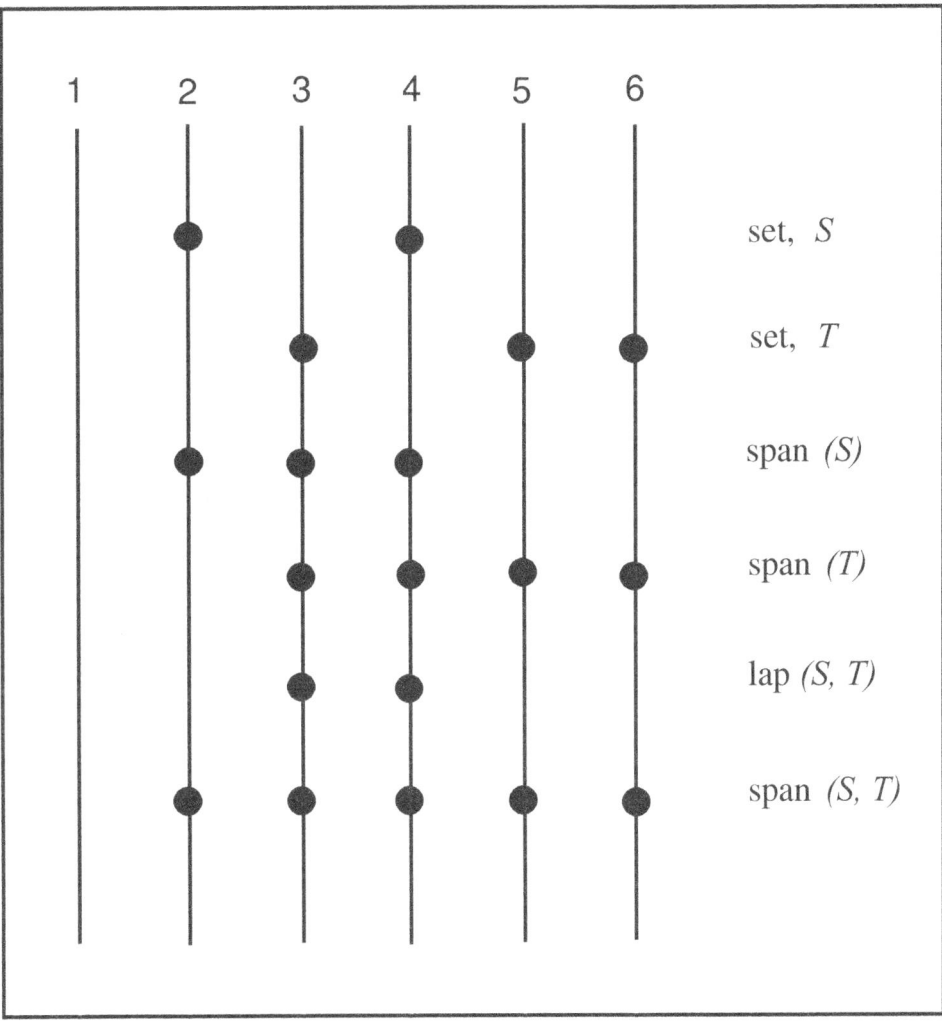

Figure 8: Computation of the weight of entanglement of two sets.

Chapter 8. The Consciousness Model

At last we apply the AR model to consciousness, and propose a solution to the mind/body problem.[122]

In Part One we have described various multi-level maps of cosmic consciousness. In this chapter we will mainly focus on two levels. Dualist-interconnectionist models for consciousness, from Ancient Greece to Descartes, have disjoint parts connected by a mysterious communication process. Usually no explanation is proposed for this communication process, although the resonance metaphor is sometimes mentioned. Here we consider this problem in the context of the mind/body model of Descartes. The intractability of this mind/body problem has been discussed from Plato on. We are going to apply to it an atomistic mechanism deriving from our models for the quantum vacuum. We thus bring together the mind/body problem of Descartes and the digital philosophy of Fredkin[123] and others[124] into a joint picture first described by Democritus.[125]

Our AR model is a process by which the illusion of continuous space self-organizes from a discrete substructure – a submicroscopic, corpuscular, dynamical cellular network – a sort of finite point set on steroids. In this chapter we further extend the AR process from space to spacetime in the domain of terrestrial physics, and then jump up to the mental and spiritual realms, where the constraints of physics no longer apply.

We apply the process twice, once to the mind, and again to the body, to obtain our resolution of the mind/body problem. In our final, composite picture, there is one enormous point set,[126]

---

[122] This chapter is adapted from our second joint paper, (Abraham and Roy, 2007).
[123] Fredkin, 2000)
[124] (Hey, 1999)
[125] (Popper, 1998)
[126] (Its size has been estimated by Wheeler as 10 to the power 88 (Hey,

operating beneath the perceived realities of macroscopic mind, body, and also quantum reality.

We begin with a review of the Mind/Body problem.

8.1. The Mind/Body Problem

The mind/body problem is a perennial thread in philosophy, East and West, so there are many illustrious names on its chronology. We will concentrate on just a few of these, to establish the main milestones of our story, and briefly describe their contributions. The earliest history, beginning with Homer, has been described by Jaspers.[127]

Plato, 370 BC

Plato's theory of the soul is described in Chapter One. In sum, we have from Plato a monistic four-level, hierarchical cosmology, comprising (from the top):

1. The Good, an integral principle with no spatial extent,
2. The Intellect, including the Ideas or Forms,
3. The World Soul (including individual human souls), and
4. The Terrestial Sphere of matter and energy.

This is very similar to the four hypostases of Plotinus. Forms exist in the Intellect, and are outside of space and time. Terrestrial objects are instances, or particulars, of Forms. Individual souls are pieces of the World Soul which have instantiated, or

---
1999).

[127]See Jaspers (1998, Essay 8).

incarnated, a Form. When people die, their individual souls reunite with their Forms.

To this Theory of Forms, Plato himself raised an objection, in his dialogue, Parmenides. This problem, later called the third man argument, or TMA, has been the subject of much discussion over the past fifty years. It is somewhat like the Russell paradox of mathematical set theory. That is, if a Form (a class of objects) contains itself as a member, then an unwelcome infinite regress is set up, toward larger and larger collections.

Some have interpreted this objection another way, which we shall call TMA2. This applies when we have two categories which are disjoint – such as two parallel universes – and yet which exchange information. A matrix between the two categories – such as the air between two resonant guitar strings – must be interpolated, to carry the resonance or intercommunication. For example, in Plato's cosmology, the World Soul intervenes between the Intellect and the Terrestrial Sphere. Or on the individual level, Ficino's Spirit intervenes between the individual soul and the body.

All this may be regarded as the prehistory of the mind/body problem.

Kashmiri Shaivism, 1000 CE

The Indian tradition provides a number of different schemes for levels of consciousness, including five koshas, seven chakras, thirty-six tattvas, and so on. The five koshas are, from the top down: the bliss body (anandamaya kosha), astral body (vijnanamaya kosha), mental body (manomaya kosha) pranic body (pranamaya kosha), and the food body (annamaya kosha). The bliss body is described as an experience of total transcendence, where only the fundamental vibration of the unconscious system remains.[128] The thirty-six tattvas have been described in Chapter 2.

The TMA2 problem may be the ultimate cause of the profusion of levels in the Sanskrit literature on consciousness. No matter how many levels, the mystery of the communication between adjacent levels in the hierarchy remains. The vibration metaphor addresses this mystery, but still begs an encompassing matrix or medium to carry information from level to level. The vibration metaphor entered the Indian literature in the Spanda (vibration), Urmi (wave), and Prana (life-force) concepts of the Trika philosophy described in Chapter 2.[129]

We may regard the mind/body problem as just the bottom level of a stack of similar problems. We intend that our attack on the mind/body problem should eventually be applied throughout the koshas, chakras, or tattvas of a full model of collective consciousness and unconsciousness. Like Plato, Kashmiri Shaivism is non-dual, but nevertheless, is plagued by a mind/body problem.

Descartes, 1632

AS described in Section 1.5, Descartes was a dualist, to whom the world consisted of two original substances — body and mind – between which there was an enormous gulf. Man consists of body and mind, which interact through the pineal gland. His dualist theory, and his mechanical view of nature, dominated philosophy for centuries. His method of thought and his theories have been subjected to devastating criticism.[130] For many historians, the mind/body problem in Western philosophy began with Descartes.

---

[128](Saraswati, 1998; p. 54)

[129]See (Probhananda, 2003, 2004), (Dyczkowski, 1992), and (Singh, 1980.).

[130]See, for example, (Jaspers, 1964).

8.2 The AR Model

In this section we briefly review the AR process from the preceding chapter. In the next section, we extend it from space to spacetime, and finally, we apply the process to the mind/body problem.

Recall that the AR model is a two-level system. The microscopic level, QX, is a dynamical cellular network of nodes and bonds. Inspired by the cellular automata of Ulam and von Neumann, a dynamical cellular network is a directed graph with connections (directed links, bonds) which appear, disappear, and change direction, according to dynamical rules.

The macroscopic level, ST that self-organizes from QX is an another dynamical cellular network, in which the nodes (supernodes) are the cliques of the QX level, bound into a network by superbonds. Finally, a neural network process imbeds the ST level into Euclidean spacetime, EST.

Thus, in our model, the ambient space of nature according to consensual reality, is actually an epiphenomenon of the atomistic and finite QX network, according to the scheme:

$$QX \rightarrow ST \rightarrow 3ST$$

This is the full AR process, which we call *condensation*. Actually, condensation is a two-step process: digital condensation followed by digital-to-analog (A/D) spatial embedding. Further, the digital condensation process involves collecting nodes of QX into groups (cliques, actually) which behave as supernodes of ST. So in a sense, ST contains QX, and we may imagine that the two dynamical cellular networks are entangled in a coevolutionary process: changes in ST feed back into the dynamical process of QX. We may indicate this scheme as,

$$QX \Leftrightarrow ST \rightarrow 3ST$$

8.3 The Two Time Dimensions

The discrete, microscopic time parameter, t, used above does not represent macroscopic time. Rather, we propose to obtain macroscopic spacetime through our process of condensation. Macroscopic time, T, exists locally as a function on spacetime, but we may pretend that there is a cosmic time function, to simplify the exposition. This would be a function on spacetime that assigns to each event a globally defined macroscopic time parameter, T. We propose now to obtain macroscopic spacetime from the condensation process applied repeatedly to the entire QX object, which contains all times, although it is rapidly changing.

The condensation process is regarded as being accomplished in a single instant, and it determines instantaneous states for the macrocosmic system in which space appears to be a continuum. Even so, the network, QX, is changing rapidly by a time-discrete process, with microtime t. We are going to regard the stepwise increasing network microtime as an internal process variable that is distinct from the continuous physical time aspect of the spacetime of general relativity, cosmic time, T. Thus, we envision two dimensions of time.

We adopt the mathematical perspective of general relativity, called the Cauchy process, in which the Einstein equation is regarded as an evolutionary system of partial differential equations. The Cauchy process for this system regards the past and present as known, and the future to be determined by integration of the system of equations along so-called characteristic curves. The topology of spacetime, along with the geometry (the metric tensor) and the physical parameters (energy, mass, electromagnetic fields, etc.) must evolve according to the Einstein equation. Wormholes and black holes may evolve as focal points of the characteristic curves.

Alternatively, for a mathematically less-challenging exposition, we may suppose, like Einstein, that spacetime is created as a finished system, a complete pseudo-Riemannian geometrical object.

So this is our proposal for the emergence of cosmic time. Constrained by the Einstein equation, cosmic time advances in discrete intervals, that might be multiple steps of microcosmic time, giant steps. With each giant step, yet another condensation occurs, as follows.

We consider a memory device, controlled by the cosmic-time function, T. Between cosmic time T_1 (corresponding to network time t_1) and cosmic time T_2 (with its network time t_2) the memory device records all of the finite states of QX between network-time t_1 and network-time t_2, and condenses this finite set of QX states into a spacelike continuum corresponding to the discrete cosmic time T_2. One method for the condensation of a finite set of QX states is the sum algorithm. That is, we form a QX sum-state by adding the node-states of all nodes, and the bond-states of all the bonds, of the set of QX states. In other words, fix a node of QX. Sum up the node-states of that one node for all the QX states with network time in the interval, $(t_1, t_2]$. Do likewise for each bond of QX, but round down if this sum is greater than one, and round up if less than minus one.

Thus, spacetime is squeezed from the dynamical cellular network, QX, as toothpaste from a tube. As giant steps are still very small compared with the resolving power of macroscopic science, cosmic time appears to be continuous. The macroscopic system, QX, sparkles with activity on the scale of Planck space and time, while macroscopic spacetime unrolls essentially continuously. The past and present become known, while the future remains yet a mystery.

In summary, our scheme,

$$QX \Leftrightarrow ST \to 3ST$$

is extended to the scheme

$$QX \Leftrightarrow ST \to 4ST$$

all in the context of the body, that is, the physical world. We now wish to apply this new scheme to the mind/body problem.

8.4 The Mind/Body Problem Resolved

We now consider two QX networks: QX_1 (the body level), QX_2 (the mind level). Each of them might be the basis for an AR process, one condensing to the body, or the physical world, as we have considered up to this point, the other to a separate world of the mind.

However, we may prefer alternatively to join QX_1 and QX_2 into a single entangled network, QX_*, on which two condensation processes operate. We might compare this approach to John Whitney's concept of digital harmony, in which a single mathematical algorithm is employed to compose a piece of music, and an abstract animated image, which then seem when played together to harmonize, due to deriving from a common archetypal process.[131] But we will proceed now with QX_1 and QX_2.

After all this preparation, our approach to the perennial conundrum is now simple: we apply the idea of condensation from a QX network twice: once to the body level, as in the AR model, and again by analogy to the mind level, as in Fredkin (2000). This results in the four-part scheme:

$$\begin{array}{rcl} QX_2 & \Leftrightarrow & Mind \\ \Updownarrow & & \\ QX_1 & \Leftrightarrow & Body \end{array}$$

The mystery connection between the disjoint mind and body systems now becomes an epiphenomenon of the connection between QX_1 and QX_2 which is not mysterious at all. For the

---
[131](Whitney, 1980)

nature of the QX model of AR is that of a dynamical cellular network, and we may regard QX_1 and QX_2 as combined into a single, entangled network, as directed links between the two systems will be allowed by our dynamical rules.

8.5 Summary

In sum, then, the mind/body connections are completed in a circuit outside ordinary consensual reality in a submicroscopic atomic realm beyond our senses, but revealed by the progress of modern physics. This realm or matrix, an extension of the quantum vacuum into the realm of consciousness, is a finite, discrete, digital, cosmos, which condenses – in the human perceptual and cognitive process – into epiphenomena, the continuum illusion of mind/body, hypostases, koshas, cakras, tattwas, and so on, of the perennial traditions of consciousness studies.

Note that the QX level is a static point set with a dynamic network structure, changing in microscopic time, t. Meanwhile, the macroscopic body and mind have been constructed as complete spacetime worlds, with locally defined macroscopic times, T. This provides a background for psi phenomena such as telepathy and clairvoyance, but also leaves a window of opportunity for free will. Like a zipper closing, the past is zipped (or firmed) up, while the microscopic future is subject to interaction with the macroscopic body and mind, until the zipper closure arrives, and condensation (or collapse) occurs.

The end of our construction is an echo of the Two Ways of Parmenides described in Chapter 1, the atomic QX_*, and the $4ST$ continua of body and mind, playing out in digital harmony.

Conclusion

At the beginning, in 2006, we aimed only to simplify, model, and clarify RR, the Requardt and Roy model for the Quantum Vacuum (Chapter 6 and Appendix 1) . This resulted in our derived model, the AR process, and our first joint paper (Chapter 7 and Appendix 3) in 2007. When we soon discovered our shared interest in meditation and consciousness studies, it occurred to us to apply the AR model to consciousness in the spirit of classical Sanskrit philosophy.[132] This resulted in our second joint paper, on the mind/body problem (Chapter 8 and Appendix 3) also in 2007. Finally, we decided, also in 2007, to put all this together in a book. It was during this book process , while researching the precursor literature over a two-year period, that we discovered extraordinary correlations between our model and the history of philosophy, East and West. The book evolved in new directions, and finally we wanted to connect our work to these ideas:

- the primacy of a creative source outside of space and time,
- the recovery of discrete space and time, and
- the interconnection of the extended universe.

Let's look back now at these three goals.

C.1 Monism

The Good of the Platonic tradition, (Chapter 1), or the Śiva tattva of the Trika cosmos (Chapter 2), is to be understood as the source and creator of the universe, outside of space and time. In the monist traditions of both East and West, the source precedes the creation of space and time. In our model this role is played by a monster dynamical cellular network, QX_*.

---

[132]See Appendix 2 for some background of this.

In Kashmiri Shaivism, they say that the creation of the universe in 36 tattvas was accomplished in a blink, as Śiva opened his eyes. Even so, we think of the 36 tattvas as unrolling in an ordered sequence, and the "density" as it were increasing step-by-step until, at the end, the levels of matter, energy, spacetime, and physical forces condense. Space and time are tattvas coming at the tail end of the process.

Although explicitly nondualist, these traditions seem to lack any interconnection scheme. Our model provides a dynamical cellular network that contains and connects all levels of consciousness.

C.2 Discrete Space and Time

Chapters 3 and 4 of Part One, and Chapters 6 and 7 of Part Two, provide a basis for discrete space and time all the way from ancient Greece to the latest developments of quantum physics. There seems little doubt now that discrete space and time are fundamental to our conception of the universe.

It is the argument of Steven M. Rosen, in *The Dimensions of Apeiron*, that ancient concepts of space and time were originally discrete.[133] After Plato, the continuum emerged into cultures worldwide, and this process concluded with the denial of atomism in the works of Descartes: the triumph of Maya. It is not unlikely that some of the problems of our time derive from this illusion of continuity, and thus, there is a need to revive atomistic concepts of space and time. In modern science, especially in the process physics of Whitehead, this revival is under way.[134] The demise of the continuum models began in 1839, according to Rosen.[135]

---

[133] (Rosen, 2004)
[134] See Figure 1 in Chapter 4.
[135] (Rosen, 2004; p. xxx)

Our discrete network models, from which space and time emerge from a process of condensation and smoothing, support this revival. Further, they suggest a similar model for cosmic consciousness, and a possible resolution of the mind/body problem.

C.3 The Interconnection of the Extended Universe

Supposing a model of the extended universe in several levels — for example, five koshas or the thirty-six tattvas — we propose a QX network and an AR process of condensation for each level. The interconnection or entanglement of adjacent levels is accomplished via mathematical connections (bonds between nodes). Thus there is no need to search for physical forces to connect, for example, collective consciousness and the physical universe, or the individual mind and body.

In fact, it makes no sense to propose physical forces, even quantum forces, as a mechanism connecting adjacent tattvas. This could work only under the materialist paradigm, in which a thought is a biochemical state of the physical brain, as in the models of Eccls, Sperry, Stapp, Penrose, Hameroff, et al.

Instead, we propose only mathematical connections between tattvas, that all tattvas are realizations of dynamical cellular networks, and the carriers of influence between them are mathematical. That is, dynamical links pop up between nodes on different levels of creation according to rules connect the whole shebang. Although we propose mathematical connections between each tattva and its sequel, we might focus now on the bottom of the chain. Let us image just three levels, as in Plato's cosmology below The Good. Thus,

- MX = the mental sphere, consciousness (a network)
- QX = the terrestrial sphere, the universe (a network)

- ST = the universe of matter and energy (the classical continuum)

We have proposed: $MX \to QX$ by links, and $QX \to ST$ by condensation. Of course, we could easily interpose another network, the quantum vacuum, QV, so: $MX \to QX \to QV \to ST$ or similarly, identify $QX = QV$ as in RR, hence $MX \to QV \to ST$ as in the models of Eccles, Sperry, Stapp, Penrose, and Hameroff discussed previiously.

In any case, we point out that $QV \to ST$, the quantum mechanical model for mental states, does not solve the mind/body problem. Our proposal goes to the heart of the mind/brain problem in a monistic way, consistent with the cosmologies of Plato and the Trika philosophy

In any case, what are space and time, really? We are proposing that the real spacetime is a construction (condensation) from a dynamical mathematical object "outside", hidden by Maya, and appearing to us as a submicroscopic granulation, smoothed over by the limitations of our perceptual organs.

This model easily accommodates the the measurement problem of quantum mechanics (Chapter 4 of part One), the big five paranormal phenomena (Chapter 5 of Part One), and the most important target of all, the mind/body problem (Chapter 8 of Part Two).

As the nondualists like Plato and Abhinava have been tellimg us, this mathematical consciousness contains our illusion of ordinary reality, mind contains both, and the superreal is primary — *demystifying the ākāśa.*

Appendix 1.
(Quantum) Space-Time as a Statistical Geometry of Fuzzy Lumps and the Connection with Random Metric Spaces

By Manfred Requardt [136] and Sisir Roy [137]

Abstract

We develop a kind of *pregeometry* consisting of a web of overlapping *fuzzy lumps* which internal with each other. The individual lumps are understood as certain closely entangled subgraphs (*cliques*) in a dynamically evolving network which in a certain approximation, can be visualized as a time-dependent *random graph*. This strand of ideas is merged with another one, deriving from ideas, developed some time ago by Menger et al. that is, the concept of *probabilistic-* or *random metric spaces*, representing a natural extension of the metrical continuum into a more microscopic regime. It is our general goal to find a better adapted geometric environment for the description of microphysics. In this sense one may it also view as a dynamical randomisation of the *causal-set* framework developed by e.g. Sorkin et al. In doing this we incorporate, as a perhaps new aspect, various concepts from *fuzzy set theory*.

---

[136]requardt@theorie.physik.uni-goettingen.de, Institut fur Theoretische Physik, Universitat Gottingen, Germany

[137]sisir@isical.ac.in), Indian Statistical Institute, Calcutta - 700 108, India

A1. Introduction

There exists a certain suspicion among the scientific community that nature may be discrete or rather "behaves discretely" on the Planck scale. But even if one is willing to agree with this "working philosophy", it is far from being evident what this vague metaphor actually means or how it should be implemented into a concrete and systematic inquiry concerning physics and mathematics in the Planck regime.

There are basically two overall attitudes as to "discreteness on the Planck scale", the one starts (to a greater or lesser degree) from continuum concepts (or more specifically : concepts being more or less openly inspired by them) and then tries to detect or create modes of "discrete behavior" on very fine scales, typically by imposing quantum theory in full or in parts upon the model system or framework under discussion. We call this the "top down" approach.

There are prominent and very promising candidates in this class like e.g. *string theory* or *loop quantum gravity* with e.g. *Spin networks* emanating from the latter approach. As these approaches are widely known we refrain from citing from the vast corresponding literature. We recommend instead two more recent reviews dealing with the latter approach but containing also some remarks about the former one ([1] and [2]). As a beautiful introduction to the conceptual problems of quantum gravity in general may serve e.g. [3].

In the following investigation we undertake to describe how macroscopic space-time (or rather, its underlying *mesoscopic* or *microscopic* substratum) is supposed to emerge as a *superstructure* of a *web of lumps* in a fluctuating dynamical cellular network. One may call this the "bottom up" approach. In doing this, two strands of research are joined, which, originally, started from different directions. The one is the *cellular network* and *random graph* approach, developed by one of the authors

(M.R.), the other the *statistical geometry of lumps*, a notion originally coined by Menger and coworkers and being further developed by various groups (see e.g. [17], [18], [19], [20]). It is worth mentioning that Einstein himself was not against such a grainy substratum underlying our space-time continuum (see the essay of Stachel in [21]).

The point where these different strands meet is the following. In the dynamical network approach of (M. R.) macroscopic space-time is considered to be a coarsegrained emergent phenomenon (called an *orderparameter manifold* in [8]). It is assumed to be the result of some kind of *geometric phase transition* (very much in the spirit of the physics of self-organisation). This framework was developed in quite some detail in e.g. [8]. We argued there that, what we consider to be the elementary building blocks of continuous space-time, i.e. the so-called *physical points*, are on a finer scale actually densely entangled subclusters of *nodes* and *bonds* of the underlying network or graph. In [8] we called them also *cliques* (which denote in graph theory the maximal complete subgraphs or maximal subsimplices of a given graph).

We further argued there that the substructure of our space-time manifold consists in fact of two stories, the primordial network, dubbed by us QX, and, overlying it, the web of lumps or cliques, denoted by ST, which can also be viewed as a coarser mesoscopic kind of network with the cliques or lumps as *supernodes* and with *superbonds* which connect lumps having a non-void overlap. This correspondence raises the possibility to relate the lumps or cliques of [8] with the lumps occurring in the approach of Menger et al.

One should however note that the two philosophies are not entirely the same. In [8] and related work the lumps emerge from a more primordial discrete dynamical substratum and consequently have a specific internal structure. In the approach of Menger et al. (at least as far as we can see) they figure as the not further resolvable building blocks of space-time if one approaches the so-called Planck-regime from above, i.e. from the continuum side. In other words, the former approach is more bottom-up oriented while the latter one is more top-down. About such not further resovable scales of space-time (which the ordinary continuum picture ends) was of course also speculated by quite a few other people, most notably Wheeler (see e.g. [9], p. 1203 ff.).

Our personal working philosophy is that space-time at the very bottom (i.e. near or below the notorious Planck scale) resembles or can be modeled as an evolving information processing cellular network, consisting of elementary modules (with, typically, simple internal discrete state spaces) interacting with each other via dynamical bonds which transfer the elementary pieces of information among the modes. That is, the approach shares the combinatiorial point of view in fundamental space-time physics which has been advocated by e.g. Penrose. It is a crucial and perhaps characteristic extra ingredient of our framework that the bonds (i.e. the elementary interactions) are not simply dynamical degrees of freedom (as with the nodes their internal state spaces are assumed to be simple) but can a fortiori, depending on the state of the local network environment, be switched on or off, i.e. can temporarily be active or inactive! This special ingredient of the dynamics hopefully allows the network to perform *geometric phase transitions* into a new ordered phase displaying a certain *two-story structure* to be explained below. This conjectured emergent geometric order can be viewed as kind of a discrete *proto space-time* or *pregeometry* carrying metrical, causal and dimensional structures (as to the *cellular network approach* cf. [4] to [8], the other point of view is expounded in e.g. the book of Roy; [19] to which we also refer for a more complete list of references).

We will see that several types of distance concepts do emerge in our analysis on the various scales of resolution of space-time. On the in our scheme most fundamental level, that is, the primordial network (or graph), we have a natural distance concept (node distance). The same holds for the network of lumps, derived from it (the *clique graph* : see below). If we leave these fundamental levels (of relatively discrete behavior) and enter the realm of quasi-continuity, other possibilities do emerge. It is interesting to relate these various concepts to each other and to distance concepts discussed e.g. in [20] or [40]. We show in particular that our network of fuzzy lumps leads naturally to the concept of *random metric spaces* (see section (7.2)). Spaces of this kind should emerge at the interface between the presumed discrete Planck scale scenario and a perhaps *quasi-continuous*, that is, coarse-grained spatial environment, connecting this primordial level with the ordinary continuous regime.

Our overall working philosophy is that geometry has to emerge from a purely *relational* picture. We hence embark on a reconstruction of the continuum concepts of ordinary physics and/or mathematics from more primordial *pregeometric* (and basically discrete) ones which, in our view, hold sway on the Planck scale. We presume that this may also shed some light on the better understanding of quantum theory as such and on the various top-down approaches mentioned above. We note in passing, that this point of view has a venerable history of its own, beginning with e.g. Leibniz. A perhaps related but technically different approach was developed by Isham et al. some time ago (see e.g. [22] or [23]), who chose to quantise metrical concepts (a more recent approach is [24]). It would be an interesting task to relate these two approaches to each other.

As another precursor may be considered the *causal-set* approach of Sorkin et al. see e.g. [31] or [32]). In a sense our framework adds dynamics, viz. interaction, and random behavior to this picture. subsection*A2. The Cellular Network Environment Motivated by the above working philosophy, we emulate the underlying substratum of our world, or, more specifically, of our space-time (quantum) vacuum (containing however in addition all the existing quantum and macro objects as extended excitation patterns!) by what we call a cellular network. This discrete structure consists of elementary *nodes*, n_i, which interact (or exchange information) with each other via *bonds*, b_{ik}, playing the role of (in this context) not further reducible elementary interactions. The possible internal structure of the nodes (modules) or bonds (interaction channels) is emulated by discrete internal state spaces carried by the nodes/bonds. The node set is assumed to be large but finite or countable. The bond b_{ik} is assumed to connect the nodes n_i, n_k. the internal states of the nodes/bond are denoted by s_i, J_{ik} respectively. As our philosophy is, to generate complex behavior out of simple models we, typically, make simple choices for them, one being e.g.

$$s_i \in q \cdot Z, J_{ik} \in \{-1, 0, +1\} \tag{1}$$

with q an elementary quantum of information.

As in our approach the bond states are dynamical degrees of freedom which, a fortiori, can be switched off or on, the *wiring*, that is the pure *geometry* of the network is also an emergent, dynamical property and is *not* given in advance. Consequently, the nodes and bonds are typically not arranged in a more or less regular array, a lattice say, with a fixed nea-/far-order. This implies that *geometry* will become to some extent a *relational* (Machian) concept and is no longer an *ideal element* (cf. the more detailed discussion in [12], which deals primarily with the emergence of quantum theory as a consequence of the geometric fine structure of such a network).

On the other side, as in cellular automate, the node and bond states are updated (for convenience) in discrete clock time steps, $t = z \cdot \tau, z \in Z$ and τ being an elementary clock time interval. This updating is given by some *local* dynamical law (examples given below). In this context *local* means that the node/bond states are changed at each clock time step according to a prescription with input the overall state of a certain neighborhood (in some topology) of the node/bone under discussion. We want however to emphasize that t is *not* to be confounded with some *physical time*, which, for its part, is also considered to be an emergent coarse grained quantity. The well known *problem of time* is, for the time being, not treated in detail in the following, as it presents a big problem of its own, needing a careful and separate analysis (see e.g. [10] or [11]). That is, at the moment the above clock time is neither considered to be dynamical nor observer dependent. We discussed however the presumed emergence of a new primordial time scale which sets the scale for the regime where quantum fluctuations hold sway in [12].

A simple example of such a local dynamical law we are having in mind is given in the following definition.

Definition 2.1 (Example of a Local Law) *At each clock time step a certain quantum q is exchanged between, say, the nodes n_i, n_k, connected by the bond b_{ik} such that*

$$s_i(t+\tau) - s_i(t) = q \cdot \sum_k J_{ki}(t) \qquad (2)$$

(i.e. if $J_{ki} = +1$ a quantum q flows from n_k to n_i etc.)

The second part of the law describes the back reaction *on the bonds (and is, typically, more subtle.) Thus is the place where the so-called 'hysteresis interval' enters the stage. We assume the existence of two 'critical parameters' $0 \leq \lambda_1 \leq \lambda_2$ with :*

$$J_{ik}(t+\tau) = 0 \; if \; \mid s_i(t) - s_k(t) \mid =: \mid s_{ik}(t) \mid > \lambda_2 \qquad (3)$$

$$J_{ik}(t+\tau) = \pm 1 \; if \; 0 < \pm s_{ik}(t) < \lambda_1 \qquad (4)$$

with the special proviso that

$$J_{ik}(t+\tau) = J_{ik}(t) \text{ if } s_{ik}(t) = 0 \qquad (5)$$

On the other side

$$J_{ik}(t+\tau) = \begin{cases} \pm 1 & J_{ik}(t) \neq 0 \\ 0 & J_{ik}(t) = 0 \end{cases} \text{ if } \lambda_1 \leq \pm s_{ik}(t) \leq \lambda_2 \qquad (6)$$

In other words, bonds are switched off if local spatial charge fluctuations are too large, switched on again if they are too small, their orientation following the sign of local charge differences, or remain inactive.

Another interesting law arises if one exchanges the role of λ_1 and λ_2 in the above law, that is, bonds are switched off if the local node fluctuations are too small and are switched on again if they exceed λ_2. We emulated all these laws on a computer and studied a lot of network properties. The latter law has the peculiar feature that it turned out to have very short transients in the simulations, i.e. it reaches an attractor in a very short clock time. Furthermore these attractors or state-cycles turned out to be very regular, that is, they had a very short period of typically six, that is, the whole network returned in a previous state after only six clock time steps, which is quite remarkable, given the seeming complexity of the evolution and the huge phase space([13]).

Remarks :

1. It is important that, generically, such laws, as introduced above, do not lead to a reversible time evolution, i.e. there will typically exist *attractors* or *state-cycles* in total phase

space (the overall configuration space of the node and bond states). On the other hand, there exist strategies (in the context of cellular automata!) to design particular *reversible* network laws (cf. e.g [14]) which are, however, typically of second order. Usually the existence of attractors is considered to be important for *pattern formation.* On the other side, it may suffice that the phase space, occupied by the system, shrinks in the course of evolution, that is, that one has a flow into smaller subvolumes of phase space.

2. In the above class of laws a direct bond-bond interaction is not yet implemented. We are prepared to incorporate such a (possibly important) contribution in a next step if it turns out to be necessary. In any case there are not so many ways to do this in a sensible way. Stated differently, the class of possible physically sensible interactions is perhaps not so numerous.

3. As in the definition of evolution laws of *spin networks* by e.g. Markopoulou, Smolin and Borissov (see [15] or [16]), there are in our case more or less two possibilities : treating evolution laws within an integrated space-time formalism or regard the network as representing space alone with the time evolution being implanted via some extra principle (which is the way we have chosen above). The interrelation of these various approaches and frameworks, while being very interesting, is however far from obvious at the moment and needs a separate detailed investigation.

Observation 2.2 (Gauge Invariance) *The above dynamical law depends nowhere on the absolute values of the node charges but only on their relative differences. By the same token, charge is nowhere created or destroyed. We have*

$$\Delta(\sum_{QX} s(n)) = 0 \qquad (7)$$

To avoid artificial ambiguities we can e.g. choose a fixed reference level and take as initial condition respectively constraint

$$\sum_{QX} s(n) = 0 \qquad (8)$$

There are many different aspects of our class of cellular networks one can study in this context. One can e.g. regard them as complex dynamical systems, or one can undertake to develop a statistical or stochastic framework etc. In a purely geometric sense, however, they are evolving *graphs*. As we are in this paper primarily concerned with the analysis of the *microstructure* of (quantum) space-time, it seems to be a sensible strategy to supress, at least in a first step, all the other features like e.g. the details of the internal state spaces of nodes and bonds and concentrate instead on their pure *wiring diagram* and its *reduced (graph) dynamics*. This is already an interesting characteristic of the network (perhaps somewhat reminiscent of the *Poincaré map* in the theory of chaotic systems) as bonds can be switched on and off in the course of clock time so that already the wiring diagram will constantly change. Furthermore, as we will see, it encodes the complete *near-* and *far-order structure* of the network, that is, it tells us which regions are experienced as near by or far away (in a variety of possible physical ways such as strength of correlations or with respect to some other physically meaningful metric like e.g. *statistical distance* etc.). Evidently this is one of the crucial features we expect from something like physical space-time. In the above simple scenario with $J_{ik} = \pm 1$ or 0 one can e.g. draw a *directed bond*, d_{ik}, if $J_{ik} = +1$, with $J_{ik} = -J_{ki}$ implied, and delete the bond if $J_{ik} = 0$. This leads to a (clock) time dependent graph, $G(t)$, or *wiring diagram*. In other words, we will deal in the following mainly with the evolution and structure of large *dynamical* graphs.

We close this section with a brief résumé of the characteristics an interesting network dynamics should encode (in our view).

Résumé 2.3 *Irrespectively of the technical details of the dynamical evolution law under discussion it should emulate the following, in our view crucial, principles, in order to match certain fundamental requirements concerning the capability of* emergent *and* complex *behavior.*

1. *As is the case with, say, gauge theory or general relativity, our evolution law on the surmised primordial level should implement the mutual interaction of two fundamental substructures, put sloppily* : *"geometry" acting on "matter" and vice versa, where in our context "geometry" is assumed to correspond in a loose sense with the local and/or global bond states and "matter" with the structure of the node states.*

2. *By the same token the alluded* selfreferential *dynamical circuitry of mutual interactions is expected to favor a kind of* undulating behavior *or* selfexcitation *above a return to some uninteresting* 'equilibrium state' *as is frequently the case in systems consisting of a single component which directly acts back on itself. This propensity for the* 'autonomous' *generation of undulation patterns is in our view an essential prerequisite for some form of "*protoquantum behavior*" we hope to recover on some coarse grained and less primordial level of the network dynamics.*

3. *In the same sense we expect the overall pattern of switched-on and-off bonds to generate a kind of "*protogravity*".*

A3. The Network of Lumps

In [8] we argued that our microscopic cellular network, QX, may be capable of performing a *geometric phase transition* into a more ordered phase, dubbed QX/ST, with elementary building blocks the maximal connected subgraphs or *cliques* (maximal subsimplices) occurring in the primordial graph, belonging to QX. We further argued that this emergent *orderparameter manifold* (as it plays a role similar to ordinary orderparameters in the statistical mechanics of phase transitions) may constitute a stochastic protoform of our ordinary space-time or quantum vacuum with the cliques as protoforms of physical points (lumps), having an inner dynamical structure and being entangled with the other (overlapping) lumps via common nodes or bonds.

The stochastic aspects are brought in by the underlying dynamical network law, which induces, among other things, a certain amount of creation and annihilation of bonds among the microscopic nodes. As a consequence the size and shape of the cliques or lumps fluctuates in the course of network evolution. These aspects have been analyzed in quite some detail in [8] within the framework of *random graphs*.

This derived coarser network, viz. the *clique graph* or *web of lumps*, is defined as follows. The cliques are represented by new *meta-nodes*, the respective *metabonds* represent overlap of cliques (in form of common nodes).

Observation 3.1 *Note that, while this new network may be regarded as being coarser in some sense, it may nevertheless in general consist of much more nodes and bonds than the underlying primordial network. Usually there are much more maximal subsimplices than primordial nodes, as a given node will typically belong to quite a few different maximal subsimplices (cf. the estimates in [8]). This array of intersecting maximal subsimplices has the natural structure of a simplicial complex with the smaller simplices as faces of the maximal ones, viz. the cliques (as to this notion see any textbook on algebraic topology like e.g. [25], cf. also section 3.2 of [6]). If we represent this simplicial complex by a new (clique-) graph with only the maximal simplices occurring as meta-nodes we loose, on the other side, some information, as we do not keep track of, to give an example, situations where, say, three lumps or cliques have a common overlap. This situa-*

tion is not distinguished from the scenario where only each pair of the triple has a common overlap.

Remark 3.2 *On the other hand, it may be physically advantageous to loose microscopic information in a controlled way. We will show below that this possibility to consider the network as a simplicial complex stops at this first level, the coarser picture of a clique graph, however, allows on the other side for a geometric renormalization group procedure as we can repeat this process on each level, thus creating a whole tower of such metagraphs, one lying over the other, leading in the end to webs which resemble more and more our ordinary space-time.*

We now have to make the physically motivated assumption that our network of lumps, which is assumed to be in the *phase QX/ST*, is not fluctuating too wildly, put differently, that the effective dynamics, induced by the underlying microscopic dynamics defined above, leaves the individual cliques. C_i, sufficiently stable. We hence assume that they do not change their form too much from clocktime step to clocktime step, so that we are able to keep track of the *history* of the individual cliques over an appreciable amount of time steps.

$$C_i(t) \cap C_i(t+1) \approx C_i(t) \qquad (9)$$

This means that only a small portion of nodes enters or leaves each of the cliques in every time step. We make this assumption so as to be able to perform some sort of *assemble averages* over fluctuating but individual (that is, labelled) cliques in the next section and identify them with *fuzzy lumps*.

A4. Cliques as Fuzzy Sets

We now concentrate on the (clock-time) evolution of an arbitrary (*generic*) but fixed clique, denoting its time sequence by

$$C(t_0), C(t_1), \cdots, C(t_N) \qquad (10)$$

We argued above that the clique under observation changes its shape and size only mildly from time step to time step. As it is frequently done in physics, we want to replace the above time series of a generic clique by an ensemble. To this end we define

$$\overline{C} := \bigcup C(t_i) \quad , \quad \underline{C} := \bigcap C(t_i) \qquad (11)$$

We now define a so-called *membership function*, $m(x)$, for an arbitrary node, x, in the graph, G. Traditionally, the usual habit in *fuzzy set theory* is it, to speak of *fuzzy sets*, but actually they do exist only via their membership functions (see below). That is, we relate to each cluster of sharp cliques, defined above, a so-called *fuzzy clique*, defined by its membership function. We denote the *fuzzy clique*, belonging to the above ensemble of sharp cliques by $\overline{C}$ and define its membership function, $m_{\overline{C}}$, as follows :

Definition 4.1 (Fuzzy Clique) *For each node, x, we define*

$$m_{\overline{C}} := \begin{cases} 0 & x \notin \overline{C} \\ 0 \leq p \leq 1 & x \in \overline{C} \\ 1 & x \in \underline{C} \end{cases} \qquad (12)$$

with p denoting the relative frequency of occurrence of node x in $\cup C(t_i)$, $\cup$ the disjoint union, Viz.

$$p := N(x)/N \qquad (13)$$

$N(x)$ the number of occurrences of x in $\cup C(t_i)$.

Remark : We want to remind the reader of the assumptions we have made above. Our fuzzy clique or fuzzy lump is a set with the individual nodes carrying a certain weight, viz. its degree of membership with respect to the fuzzy clique. $\overline{C}$, which, on its side, is given by the corresponding membership function, $m_{\overline{C}}(x)$. Note that the underlying philosophy is quite modern as it replaces an underlying space (a set of points) by a class of functions on the space. A similar philosophy is e.g. hold in *non-commutative geometry* and related fields.

We should add at this place some remarks of precaution. We studied in quite some detail in [8] the statistical distribution of the number and size of occurring cliques in, say, a *random graph*. It turns out that most of the cliques have a typical (generic) size. On the other hand there exists a (possibly small) fraction of degenerated, i.e. very small ones. These small ones may even happen to vanish and/or emerge in the envisaged clock-time interval. That, is, we expect our general picture to be true modulo some stochastic noise, which we choose to neglect at the moment.

Conclusion 4.2 *Under the assumption, being made, we can, in a certain approximation, replace the clique graph introduced above by a net of overlapping fuzzy lumps. Mathematically this net is implemented by the corresponding class of membership functions, $\{m_i(x)\}$, with i labelling the fuzzy lumps and where the overlap of cliques is now encoded in the overlap of the supports of the functions, $m_i(x)$, viz.*

$$x \in \tilde{C}_i \quad \text{if} \quad m_i(x) > 0 \tag{14}$$

We see from the above that, to put it briefly, fuzzy-set theory consists of replacing the ground set, $\mathcal{P}(X)$, X some space of points, $\mathcal{P}(X)$ the set of subsets of X, and the respective operations on it, by the set of functions, I^X, I the unit interval, and corresponding operations on it. $\mathcal{P}(X)$ can be made into a *Boolean algebra* or a *lattice* with the help of the operations $\cap, \cup$ or $\cap$ as multiplication and the symmetric difference

$$\Delta(A, B) := (A \backslash B) \cup (B \backslash A) \tag{15}$$

as addition.

Corresponding relations can be given for fuzzy sets, but there are a lot of different possibilities to implement or mimik the above set-theoretic operations. The perhaps most straightforward ones are

$$\cap \to \min(\tilde{A}(x), \tilde{B}(x)) \quad \cup \to \max(\tilde{A}(x), \tilde{B}(x)) \tag{16}$$

where, for brevity, we use from now on the same symbol for the fuzzy set and its membership function. Some more remarks on these matters can be found in the following section. (A nice introduction to fuzzy set theory on a mathematically satisfying level is e.g. [26], including a huge bibliography. A brief but concise representation can also be found in [27]). Some slightly more advanced relations between fuzzy sets, being of potential use in our geometric enterprise are the "degree of being a subset" and the "degree of similarity" between two fuzzy sets. Defining the *size* of the fuzzy set by

$$[\tilde{A}] := \sum_x \tilde{A}(x) \quad \text{or} \quad \int \tilde{A}(x) d^n x \qquad (17)$$

(existence of the rhs being assumed), we define e.g. the degree of being a subset as

$$\mathrm{sub}(\tilde{B}; \tilde{A}) := [\tilde{A} \wedge \tilde{B}]/[\tilde{A}] \qquad (18)$$

and the degree of similarity of two sets by

$$\mathrm{sub}(\tilde{A}, \tilde{B}) := [\tilde{A} \wedge \tilde{B}]/[\tilde{A} \vee \tilde{B}] \qquad (19)$$

with extending the notions of ∩, ∪; one may take e.g. the above (min, max)- implementations.

The above binary relations, given e.g. by (min, max), fullfil the criteria of so-called (t, s)-norms (see the next section). There are other such functions in use, which have perhaps nicer properties. One is the so-called *Lukasiewicz-(t, s)-norm* :

$$t_L(\tilde{A}, \tilde{B}) := \max\{0, \tilde{A} + \tilde{B} - 1\} \quad s_L(\tilde{A}, \tilde{B}) := \min\{1, \tilde{A} + \tilde{B}\} \qquad (20)$$

leading to (for more details see e.g. [28]). The advantage of the Lukasiewicz-operations is that they induce a metric on fuzzy-sets. We state without proof.

Observation 4.3 $1 - \mathrm{sim}_L(\tilde{A}, \tilde{B})$ *defines a metric on fuzzy-sets.*

Remark 4.4 *We rediscover exactly this t-norm in section (7.2) in quite a different context (E-spaces).*

As to this observation, note that for ordinary (finite) sets, $\Delta(A,B)$ induces also a metric.

Observation 4.5 (Hamming-distance) $|\Delta(A,B)|$ *defines a metric on finite sets.*

Proof : The only non-trivial property is the *triangle inequality*. As a direct approach is perhaps a little bit tedious, we give a perhaps more pedagogical proof using graph theory. Subsets of a finite set, X, can be represented by functions $f \in \{0,1\}^X$. These functions, on the other side, can be represented as the vertices of a hypercube, $Q^{|X|}$. $|\Delta(A,B)|$ is now the minimal length of a path between f_A, f_B on $Q^{|X|}$, i.e., the minimal number of flips of $(0,1)$ along a path connecting f_A and f_B. This distance on graphs is however a metric. □

After this short digression we want to return to our original enterprise, i.e. to relate our approach to the *probabilistic-metric-*approach, mentioned above. We started from an underlying graph, $G(t)$, and a superimposed clique-graph. $G_{cl}(t)$, both carrying a natural distance function, i.e.

$$d(n_i, n_k) \quad , \quad d_C(C_i, C_k) \qquad (21)$$

that is, the minimal length of a path, connecting the given nodes (cf. [4] to [8]). Keeping the labelled nodes or cliques (supernodes) fixed, both these distances fluctuate in the course of clock-time evolution as bonds are switched on or off according to one of the microscopic dynamical laws, given above. The clique-metric will fluctuate since the cliques change their shape and size, viz. also their degree of overlap. On the other side, its fluctuations are supposed to be less erratic as a non-void overlap means usually that several nodes and bonds are involved, hence clique-distance should be more stable.

When we switch from this dynamical picture of a time-dependent graph, $G(t)$, to the ensemble picture of fuzzy cliques or lumps, our point of view changes to a static but, on the other side, probabilistic one. This latter point of view is more in the spirit of the framework of Menger et al. That is, the structure of the space under study is no longer time dependent while its largely hidden dynamics is now encoded in various probabilistic notions like e.g. a random metric (which we would like to derive from some underlying principles). We will make the connection to this other framework after the following sections, introducing some technical material about fuzzy-set theory and probabilistic metric spaces. Furthermore we plan to relate in this final section our ap0proach to the complex of ideas developed by e.g. Sorkin et al. (see e.g. [31] or [32]; we rocomment in particular the latter reference as a thoughtful introduction into this particular bundle of ideas).

A5. Some Concepts from Fuzzy Set Theory

We give in this section a brief introduction into the ideas of fuzzy set theory. Let A be a fuzzy set on X. Then by definition $A(x)$ is interpreted as the degree to which x belongs to A. The fuzzy operations like fuzzy unions, intersections and complements are certain (context dependent) generalizations of the corresponding classical set operations. Let e.g. B be another fuzzy set and A be the complement of A. A particular variant of such fuzzy operations are the following :

$$A(x) = 1 - A(x) \tag{22}$$

$$(A \cap B)(x) = \min[A(x), B(c)] \qquad (23)$$

and

$$(A \cap B)(x) = \max[A(x), B(c)] \qquad (24)$$

In the literature, fuzzy intersections and unions are in general defined via so-called *t-norms (triangular norms)* and *t-conorms* respectively. However, it should be noted that such fuzzy operations are by no means unique. Different choices of such *t-norms* may be appropriate to represent these operations in varying contexts and hence they are known as context dependent operations. Essentially fuzzy set theory provides us with an intuitive notion of uncertainty. Subsequent to the development of fuzzy set theory, *fuzzy measure theory* was developed (see e.g. [29]). The concept of fuzzy measure theory is conceptually an important step in understanding the foundational issues related to fuzzy set theory. It provides a broader framework which allows to introduce something like *possibility theory*.

Crucial in this field as well as in the field of probabilistic metric spaces is the above mentioned notion of *t-norm* (more about the history of this important concept can be found in the book of Schweizer and Sklar, [20]). The (generalized) intersection of two fuzzy sets A and B is defined by a binary operation on the unit interval as

$$t : [0, 1] \times [0, 1] \to [0, 1] \qquad (25)$$

Definition 5.1 (T-Norm)

$$t(a, 1) = a \quad (boundary\ condition) \qquad (26)$$

$$b \leq d\ implies\ t(a, b) \leq t(a, d) \quad (monotonicity) \qquad (27)$$

$$t(a, b) = t(b, d) \quad (commutativity) \qquad (28)$$

$$t(a, t(b, d)) = t(t(a, b), d) \quad (associativty) \qquad (29)$$

It is instructive to compare ordinary probability theory with *possibility theory* and in particular, probability theory with fuzzy set theory. In some sense a fuzzy measure is the dual concept to the concept of fuzzy sets. It encodes the possibility of a given fixed (fuzzy) object to belong to the respective sets $\mathcal{C}$ (cf. also [30]).

Definition 5.2 *Given a set X and a nonempty family $\mathcal{C}$ of subsets of X, which, for convenience, we take to ba a σ-algebra. A fuzzy measure on $(X, \mathcal{C})$ is a function $g : \mathcal{C} \to [0, 1]$ that satisfies the following requirements :*

1. $g(\emptyset) = 0$ and $g(X) = 1$ *(boundary requirements)*.

2. *for all $A, B \in \mathcal{C}$, if $A \subseteq B$, then $g(A) \leq g(B)$ (monotonicity)*

3. *for any increasing sequence, $A_1 \subset A_2 \subset \ldots$ in $\mathcal{C}$ so that $\bigcup_{i=1}^{\infty} A_i \in \mathcal{C}$ we have $\lim_{i \to \infty} g(\bigcup_{i=1}^{\infty} A_i)$ (continuity from below)*

4. *for any decreasing sequence $A_1 \supset A_2 \supset \ldots$ in $\mathcal{C}$, $\lim_{i \to \infty} g(A_i) = g(\bigcap_{i=1}^{\infty})$ (continuity from above)*.

In case of a probability measure we have for $A \cap B = \theta$:

$$P(A \cup B) = P(A) + P(B) \tag{30}$$

From this one sees that ordinary probability measures are a true-subclass of possibility or fuzzy measures. As we do not use these more advanced concepts at the moment, we refrain from giving more details which can be found in the mentioned literature.

A6. Concepts from the Theory of Probabilistic

In the statistical geometry, developed by Menger et al. points are no longer considered as the elementary building blocks. In some sense lumps play now their role as primordial, not further resolvable elements. Various concepts of a probabilistic nature are introduced which allow to quantify varying degrees of *(in) distinguishability* of objects. In this way Menger solved Poincare's dilemma of having, on the one side, a *transitive* mathematical and, on the other side, a possible intransitive physical relation of equality (cf. [41] and further remarks in section (7)). In this geometrical framework we have tow basic ingredients :

1. The concept of *hazy* or *fuzzy lumps*

2. The "randomisation" of various geometrical or metrical concepts

Frechet ([33]) gave an abstract formulation of the notion of distance in 1906. Hausdorff ([34]) proposed the name *metric space* and introduced the function d that assigns a nonnegetive real number $d(p, q)$ (the distance between p and q) to every pair (p, q) of elements (points) of the set S. Its properties are

$$d(p,q) = 0 \leftrightarrow p = q \qquad (31)$$

$$d(p,q) = d(q,p) \qquad (32)$$

$$d(p,r) \leq d(p,q) + d(q,r) \qquad (33)$$

In 1942 Menger ([35]), guided by the experimental situation in the natural sciences, proposed to replace the "deterministic" function $d(p,q)$ by a more probabilistic concept. He introduced the probability distribution. F_{pq}, whose value $F_{pq}(x)$ for any real number x, is interpreted as the probability that the distance between p and q is less than x.

Remark 6.1 *We assume that F_{pq} is continuous from the left. On the other side, we may nevertheless encounter the situation that $F_{pq}^{+}(0) > 0$. This means that there may be a non-vanishing probability for two different lumps to have a vanishing distance. This particular point is discussed by Menger in [40] and implements Poincare's observation of a possible non-transitive behavior.*

Since probabilities can neither be negative not be greater than 1, we have

$$0 \leq F_{pq} \leq 1 \qquad (34)$$

for any real x. Menger defined a *statistical metric space* as a set S with an associated set of probability distribution functions F_{pq} which satisfy the following conditions.

$$F_{pq}(0) = 0 \tag{35}$$

$$\text{If } p = q \text{ then } F_{pq}(x) = 1 \text{ for all } x > 0 \tag{36}$$

$$\text{If } p \neq q \text{ then } F_{pq}(x) < 1 \text{ for some } x > 0 \tag{37}$$

and

$$F_{pq}(x+y) \geq T(F_{pr}(x), F_{qr}(y)) \tag{38}$$

for all p, q, r in S and all real numbers x, y. Here T is a function from the closed unit square $[0, 1] \times [0, 1]$ into the closed unit interval $[0, 1]$ (called a *triangular function* or *triangular norm*, see below).

One may view these probabilistic metric spaces as being derived from an underlying fully probabilistic model system, living on a probability space and with the physical observables being random variables. Spacek [36] was the first to look at the subject from this point of view. He proposed the name *random metric space* instead of probabilistic metric space and discussed the relationship between these two notions. Stevens [37] in his doctoral dissertation tried to modify Spacek's approach. The main idea behind Steven's approach lies in the fact that one has a set S and a collection P of measuring rods. One chooses a measuring rod d from P at random to measure the distance between two given points, p and q, of S. With the help of this idea Stevens defined the distribution function F_{pq} and showed that the metrically generated space so obtained is a Menger space.

Menger, on the other side, started with a probability distribution function instead of random variables. This is related to the fact that the outcome of any series of measurements of the values of a nondeterministic quantity is a distribution function and the probability space may be unobservable in principle. This point of view has its roots in the positivistic philosophy (ef. e.g. [40]) and is in line with a possible nonclassical behaviour.

Sherwood [38] approached the problem from a different point of view. Following the concept of distribution-generated spaces as introduced by Schweizer and Sklar ([39]). Sherwood proposed the concept of E-space. In an E-space, the points are functions from a probability space $(\Omega, \mathcal{A}, P)$ into a metric space (M, d). For each pair, (p, q), of functions in the space, the function $d(p, q)$, defined as

$$(d(p,q))(\omega) = d(p(\omega), q(\omega)) \qquad (39)$$

for all ω in Ω is a random variable on $(\Omega, \mathcal{A}, P)$. The function F_{pq} is then the distribution function of this random variable, i.e. we have

$$F_{pq}(x) = P((d(p,q))(\omega) < x) \qquad (40)$$

In this way, F_{pq} can be regarded as the probability that the distance between p and q is less than x. Sherwood showed that every E-space is a Menger space (cf. the subsection (7.2) below).

Another subclass of probabilistic metric spaces are the so-called distribution generated spaces. The main idea is roughly as follows : Let S be a set. With each point p of S associate an n-dimensional distribution function G_p and with each pair, (p, q), a 2n-distribution function H_{pq} so that it holds :

$$H_{pq}(\vec{u}\ \vec{v} = (\infty, \ldots\ldots, \infty)) = G_p(\vec{u}) \qquad (41)$$

$$H_{pq}(\vec{u} = (\infty, \ldots\ldots, \infty), \vec{v}) = G_p(\vec{v}) \tag{42}$$

for any $\vec{u} = (u_1, \ldots\ldots, u_n)$ and $\vec{v} = (v_1, \ldots\ldots, v_n)$ in R^n.

Let now $Z(x)$ be a cylinder in R^n with

$$Z(x) := \left\{ (\vec{u}, \vec{v}) \in R^{2n}; |\vec{u} - \vec{v}| < x \right\} \tag{43}$$

for any $x > 0$. This defines a distance distribution function. F_{pq}. via

$$F_{pq}(x) = \int_{Z(x)} dH_{pq} =: P_{H_{pq}}(Z(x)) \tag{44}$$

The technical details can be found in e.g. [20]. Note that, in general, the pairs of S need not be independently distributed.

In one interpretation one may view the elements of S as "particles". Then for any Borel set A in R^n, the integral $\int_A dG_p$ can be interpreted as the probability that the particle p can be found in the set A and $F_{pq}(x)$ as the probability that the distance between the particles q and p is less than x. This then yields another type of probabilistic metric space.

A subclass consists of models where the members in S behave independently of each other. Such spaces are called *cloud-spaces* (*C-spaces*). A function g from R^n into R^+ is an n-dimensional probability density if the function G defined on R^n by

$$G(\vec{u}) = \int_{((-\infty,\ldots\ldots,-\infty),\vec{u}))} g(\vec{v}) \, d\vec{v} \tag{45}$$

is an n-distribution function. If p is a point in a distribution-generated space over R^n such that G_p is absolutely continuous, then the corresponding density g_p of G_p may be visualized as a "cloud" in R^n - a cloud whose density at any point of R^n measures the relative likelihood of finding the particle p in the vicinity of that point.

Another, equally natural interpretation (which is perhaps more in line with our concept of *fuzzy lumps;* see section (4)) is to visualize the space as an aggregate of *clouds* or *fuzzy points*. To make life simple we may assume that to all points, p, is associated a vector, $\vec{c}_p$ in R^n so that

$$g_p(\vec{u}) = g(\vec{u} - \vec{c}_p) \qquad (46)$$

with g specially symmetric. We may then replace the points by clouds in R^n and get a notion of *homogeneity*. This model is perhaps the most straightforward extension of ordinary (Euclidean) space and resembles our network model of fuzzy lumps (viewed as fuzzy sets) if we chose to embed it into ordinary R^n.

It can be shown that by taking the convolution product of $g(\vec{u} - \vec{c}_p)$ and $g(\vec{v} - \vec{c}_q)$, one can generate a particular type of random metric space. By applying then a further amount of probabilistic machinery one can create what is called *Frechet-Minkowski-metrics, d_β* on the original space S, which may e.g. be classical Euclidean space. These metrics are related with and derived from the underlying random metric space (cloud space). The to some extent intricate calculations can be found in [20], chapter 10.

These *Frechet-Minkowski-metrics, d_β*, associated with (semihomogeneous) C-spaces over R^n have a remarkable structure. At small distances this metric is non-Euclidean and the distance between two distinct points p, q is bigger than a fixed positive constant, associated with $g(\vec{u})$. On the other hand, it becomes euclidean in the asymptotic region. It appears from the above picture that if we consider a C-space as a space of clouds (which may move around), the "haziness" of the distance between p and q, which is predominant when the clouds are close together, becomes more and more insignificant when the clouds are sufficiently far apart. In this sense *Frechet-Minkowski metrics* become asymptotically euclidean.

A7. Random Metrics on our Networks of Fuzzy Lumps

In this section we want to relate the various complexes of concepts and techniques, developed or presented above, with each other. That is, we view on the one side the (dynamically) fluctuating cliques as smeared out lumps, viz. as fuzzy sets. On the other side, we want to construct so-called *probabilistic* or *random metric spaces* over this space of fuzzy lumps.

Note that this task is however not entirely straightforward as within the mathematical framework of the latter approach the underlying space is usually treated as a more or less structureless space of simple points with the underlying possible causes of the fluctuations in the metrical distance usually not being openly discussed. This has however to be done in the concretely given model systems.

We should ad the remark that there are also different scientific philosophies behind these various points of view (cf. [20] p.17). There exists a tradition to keep the (or a) underlying concrete *probability space* out of the game and regard is as secondary, while concentrating on the class of (phenomenologically given) *distribution functions*.

On the other side, a concretely given (and, what is even better, physically motivated) microscopic probabilistic ground space provides (among other things) a common and unifying reference frame. In other words, we have a canonical deterministic metric on the underlying dynamical graph, $G(t)$ or its *clique graph*, $G_{cf}(t)$ (cf. (21)). We have argued that our cliques or lumps are fluctuating as a result of the imposed dynamical laws (discussed in section 2). From this input we have to infer the concept of a *probabilistic distance* between two given labelled (fuzzy) lumps. We divide our section into two subsections. In the first subsection we introduce distance concepts on the space of overlapping (static) fuzzy lumps as they are given in the context of *fuzzy set theory* as discussed in section 4. In this context fluctuations are only implicitly included and the model theory has (one may say) the status of kind of a *mean field theory*. On the other side, the metrics, we introduce in the following subsection are related in spirit to distance concepts introduced by Menger in his interesting essay [40]. More fundamental are in our view the concepts we develop in the other subsection. We construct an explicit probability space, introduce metrical distance as an explicitly given random variable, find the corresponding *triangular norm* and show that the space we get is a model of a so-called *E-space*, introduced in Section 6. These E-spaces are a subclass of Mengers probabilistic metric spaces.

7.1 Metrics on Fuzzy Lumps, the Mean Field Picture

The perhaps most immediate method to impose some metric distance on our space of fuzzy lumps is the following one. In a first step we define a distance concept for the immediate ("infinitesimal") neighborhood of a given lump and then proceed (for finite distances) by a *chain method*, viz. paste infinitesimal distances together.

We start with two fuzzy lumps, $\tilde{C}_0, \tilde{C}_1$, which overlap as fuzzy sets. We hence have (see section 4) :

$$0 \leq p(\tilde{C}_0, \tilde{C}_1) := 1 - \text{sim}(\tilde{C}_0, \tilde{C}_1) < 1 \qquad (47)$$

We have now several possibilities to proceed. We can e.g. fix a certain scale resolution and state that two fuzzy lumps are indistinguishable if

$$p(\tilde{C}_0, \tilde{C}_1) \leq \varepsilon \qquad (48)$$

This is in accord with already classical ideas of Poincaré, developed in [41], p.31f. Put differently, in the physical world we may have a *non-transitive* infinitesimal distance concept or, more generally, a non-transitive concept of *identity* so that

$$A = B \, , \, B = C \, , \, A \neq C \qquad (49)$$

or

$$d(A,B) = 0 \, , \, d(B,C) = 0 \, , \, d(A,C) \neq C \qquad (50)$$

Definition 7.1 (Distance, Variant 1) *Denoting by γ a path connecting $\tilde{C}_i$ and $\tilde{C}_j$, which consists of a chain s.t.*

$$\tilde{C}_i =: \tilde{C}_{k_0} \, , \tilde{C}_j =: \tilde{C}_{k_n} \qquad (51)$$

and

$$p(\tilde{C}_{kj}, \tilde{C}_{k_{j+1}}) \leq \varepsilon \qquad (52)$$

we define

$$\mathrm{dist}_\varepsilon(\widetilde{C}_i, \widetilde{C}_j) := \inf_\gamma n \qquad (53)$$

($\mathrm{dist}_\varepsilon := \infty$ *if there is no such chain*).

It can be easily shown that this definition fulfils the axioms of a metric. Note however that the validity of the *triangle inequality* is enforced by attributing the distance 'one' to two "indistinguishable" lumps. This is a little bit unpleasant but inescapable if we want to base our macroscopic distance concept on the concatenation of "infinitesimal" distances. On the other side, this kind of distance is probabilistic in a very restricted sense and yields only distances which are multiples of a fixed given distance element.

We can improve the situation by taking as infinitesimal distance elements the above measures of overlap, $p(\widetilde{C}_{K_j}, \widetilde{C}_{k_{j+1}})$, and define

Definition 7.2 (Distance, Variant 2)

$$d(\widetilde{C}_i, \widetilde{C}_j) := \inf_\gamma \sum p(\widetilde{C}_{k_j}, \widetilde{C}_{k_{j+1}}) \qquad (54)$$

Note that in contrast to the preceding definition now the infimum need not be taken at a path of minimal canonical length. Quite the contrary, in exceptional situations the canonical length of a path leading to a very short distance may be quite large, whereas this is not expected to be the ordinary situation. Anyway, the second distance concept is a little bit more probabilistic.

7.2 The Space of Cliques as a Probability Space and the Connection with Random Metric Spaces

We repeat briefly what we have said in preceding sections. We have assumed that we have a network of a more or less fixed number of, however, fluctuating cliques. That is, their degree of overlap may vary in the course of time. At each discrete time step during a sufficiently long time interval, i, (so that we can exploit probabilistic concepts in, at least, a certain approximation) we have a definite array of overlapping cliques, $G_{cl}(t)$. As ground set for our probability space we take this set, i.e.

$$X := \{G_{cl}(t), t \in I, [I] = N, N \text{ some sufficiently large number}\} \tag{55}$$

The simplest assumption is to assume that all configurations have the same probability, $P(G_{cl}(t)) := 1/N$, say

Remark 7.3 *On physical grounds a more sophisticated probability density may suggest itself. Our results, derived below, are however independent of the particular choice of P.*

Observation 7.4 *The above defines a discrete probability space, X, on which random variables can be introduced, their expectation value being denoted by $<>$.*

Example 7.5 *We previously introduced the membership function, $\widetilde{C}_i(x)$, of a fuzzy lump, $\widetilde{C}_i$. Defining the elementary random variable (x a node of the primordial underlying graph)*

$$\chi_{C_i}(G_{cl}(t)) = \begin{cases} 1 & x \in C_i(t) \\ 0 & else \end{cases} \tag{56}$$

we have

$$\tilde{C}_i(x) = \langle x_{C_i} \rangle =: x(\tilde{C}_i) \tag{57}$$

(cf. the section about fuzzy lumps).

In the same sense we can define the distance between two fixed lumps, C_i, C_j, as a random variable on our probability space, taking the values $d_{(C_i,C_j)}(G_{cl}(t))$. In contrast to our ordinary fuzzy lumps, all these random variables do fluctuate, the fluctuations given by

$$\langle (x_{C-i} - \langle x_{C_i} \rangle)^2 \rangle \quad \text{or} \quad \langle (d - \langle d \rangle)^2 \rangle \tag{58}$$

For simplicity reasons we write from now on the metric as $d(p, q)$, p, q denoting some cliques, and with the understanding that it represents the above random variable.

If we want to relate our approach to the probabilistic metric framework laid out in e.g. [20], we have to inspect the *transitivity properties* of our distance concept.

Observation 7.6 *With p, q, r three points (cliques), we have for the respective random variables*

$$d_{p,r} \leq d_{p,q} + d_{q,r} \tag{59}$$

as they are evaluated on a fixed network, $G_{cl}(t)$, where the ordinary triangle inequality of the canonical graph distance holds. Hence

$$\langle d_{p,r} \rangle \leq \langle d_{p,q} \rangle + \langle d_{q,r} \rangle \tag{60}$$

On the other side, probabilistic metric spaces are defined via the probability distributions, $F_{pq}(x)$, denoting the probability that the distance between p, q is smaller than x. The ordinary triangle inequality is replaced in this framework by a more complicated estimate (cf. section 6) :

$$F_{pr}(x+y) \geq T(F_{pq}, F_{qr}) \tag{61}$$

with T a *triangular norm*.

Our task consists now in determining this function T within our own stochastic framework. To this end we employ the following formula which holds for probability measures (i.e. $P(X) = 1$).

Lemma 7.7 *With A, B measurable sets in X it holds*

$$P(A \cap B) \geq P(A) + P(B) - 1 \tag{62}$$

Proof : This result has also been employed in[20], p.26. For completeness sake we repeat it here. We have

$$P(A \cap B) = P(A) + P(B) - P(A \cup B) \geq P(A) + P(B) - P(X)$$

$$= P(A) + P(B) - 1 \tag{63}$$

and therefore

$$P(A \cap B) \geq \max(P(A) + P(B) - 1, 0) \tag{64}$$

This proves the lemma. □

We have by definition :

$$F_{pr}(x+y) = P(d_{pr} < x+y) \tag{65}$$

which, for the simplest choice of P, is just the fraction of configurations, $G_{cl}(t)$, with the canonical graph metric fulfilling this inequality. As the triangle inequality holds for the canonical graph metric we have

$$P(d_{pq} < x, d_{qr} < y) \leq P(d_{pr} < x+y) \tag{66}$$

and hence, with the help of the above lemma :

$$F_{pr}(x+y) = P(d_{pr} < x+y) \geq \max(F_{pq}(x) + F_{qr}(y) - 1, 0) \tag{67}$$

Observation 7.8 *We proved that our random metric, introduced above, fulfills a triangular inequality*

$$F_{pr}(x+y) \geq W(F_{pq}(x), F_{qr}(y)) \tag{68}$$

with W the triangular norm

$$W(a,b) := \max(a+b-1, 0) \qquad (69)$$

Hence our above probability space is a Menger space.

(That W is in fact a triangular norm is easy to show and is also employed in e.g. [20]).

A special class of probabilistic metric spaces are the so-called *E-spaces* (cf. the section 6 and [20], chapt. 9). To complete our analysis we can reframe our model so that it becomes such an *E-space*. The ground space, X, is our probability space. The metric target space, M, is a new *space-time supergraph*, defined as follows :

Definition 7.9 (Space-Time Picture of Clique Graph)

$$M := \bigcup_{t \in I} G_{cl}(t) \qquad (70)$$

with the understanding that exactly the nodes $C_i(t)$ and $C_i(t+1), t, t, +1 \in I$, are connected by additional bonds. That is, $C_i(t)$ describes the orbit of the fixed but fluctuating clique, C_i, in the course of time.

We can now relate bijectively the cliques, C_i, and certain functions from X to M.

Observation 7.10 *With the class of functions, C_i,*

$$C_i(G_{cl}(t)) := C_i(t) \in M \qquad (71)$$

and their distance

$$d_{pq}(G_{cl}(t)) := d(C_p(t), C_q(t)) \tag{72}$$

the above probability space can be considered as an E-space.

References

[1] L. Smolin : "The Future of Spin Networks" in "The Geometrical Universe", Eds. S. A. Huggett, L. J. Mason, K. P. Tod, S. T. Tsou, N. M. J. Woodhouse, Oxford Univ. Pr. Oxford 1998. gr-qc/9702030

[2] C. Rovelli : "Strings, loops and others, a critical survey of the present approaches to quantum gravity", in "Gravitation and Relativity : At the turn of the millenium", Eds. N. Dadhich, J. Narlikar, Poona Univ. Pr. 1999, gr-qc/9803024

[3] C. J. Isham : "Structural Issues in Quantum Gravity", gr-qc/9510063 (Lecture given at the GR14 Conference. Florence 1995)

[4] T. Nowotny, M. Requardt : "Dimension Theory on Graphs and Networks", J. Phys. A : Math. Gen. 31(1998) 2447, hep-th/9707082

[5] T. Nowotny, M. Requardt : "Pregeometric Concepts on Graphs and Cellular Networks", invited paper J. Chaos, Solitons and Fractals 10(1999) 469, hep-th/9801199

[6] M. Requardt : "Cellular Networks as Models for Planck-Scale Physics", math-ph/0001026

[7] M. Requardt : "Spectral Analysis and Operator Theory on (Infinite) Graphs", math-ph/0001026

[8] M. Requardt : "(Quantum) Space-Time as a statistical Geometry of Lumps in Random Networks",Class. Quant. Grav. 17(2000) 2029, gr-qc/9912059

[9] C. W. Misner, K. S. Thorne, J. A. Wheeler : "Gravitation", Freeman and Comp., San Francisco 1973

[10] C. J. Isham : "Canonical Quantum Gravity and the Problem of Time", Lectures presented at the NATO Advanced Study Inst. "Recent Problems in Math. Phys.", Salamanca June 1992, gr-qc/9210011

[11] J. Butterfield, C. J. Isham : "On the Emergence of Time in Q. Gr.", gr-qe/9901024 to appear in "The Argument of Time", ed. J. Butterfield, Oxford Univ. Pr., Oxford 1999

[12] M. Requardt : "Let's Call it Nonlocal Quantum Physics", Geottingen Preprint. gr-qc/0006063

[13] Th. Nowotny : Diploma Thesis, a compilation of some of the extensive computer simulations can be found on the web site :
http://www.physik. unileipzig.de~nowotny/research.html
(commentary in german)

[14] T. Toffoli, N. Margolus : "Cellular Automation Machines", MIT Pr., Cambridge Mass. 1987

[15] F. Markopoulou, L. Smolin : "Causal Evolution of Spin Networks", Nucl. Phys. B508 (1997) 409 or gr-qc/9702025

[16] R. Borissov : "Graphical Evolution of Spin Network States", Phys. Rev. D55 (1997) 6099 or gr-qe/9606013

[17] K. Menger : in "Albert Einstein Philosopher Scientist", ed. P. A. Schilpp, 3rd edition Cambridge Univ. Pr., London 1970

[18] N. Rosen : "Statistical Geometry and Fundamental Particles", Phys. Rev. 72 (1947) 298

[19] S. Roy : "Statistical Geometry and Applications to Microphysics and Cosmology", Kluver Acad. Publ. Dordrecht 1998

[20] B. Schweizer, A. Sklar : "Probabilistic Metric Spaces", North-Holland N. Y. 1983

[21] J. Stachel : "Einstein and Quantum Mechanics" in "Conceptual Problems of Quantum Gravity", eds. A. Asthekar, J. Stachel, Einstein Studies vol. 2. Birkhäuser Boston 1991

[22] C. J. Isham, Y. Kabyshin, P. Renteln : "AQuantum Norm Theory and the Quantisation of Metric Topology", Class. Quant. Grav. 7 (1990) 1053

[23] C. J. Isham : "An Introduction to Generla Topology and Quantum Topology", Lectures presented at the Advanced Summer Institute on Physics, Geometry and Topology, Banff August 1989

[24] C. J. Isham : "Topos theory and consistent histories", Int. J. Theor. Phys. 36 (1997) 785

[25] P. Alexandroff, H. Hopf : "Topologie" (reprint), Springer N. Y. 1974

[26] R. Lowen : "Fuzzy Set Theory", Khuwer, Dordrecht 1996

[27] "I. N. Bronstein, K. A. Semendjajew" : "Taschenbuch der Mathematik", Harry Deutsch, Frankfurt 1997

[28] B. Dermant : "Fuzzy-Theorie, Vieweg, Braunschweig 1993

[29] Z. Wang, J. H. Klir : "Fuzzy Measure Theory", Plenum Pr. N. Y. 1992

[30] H. B. Bandemer, S. Gottwald : "Einfuehrung in Fuzzy-Methoden", Akademie Verlag, Berlin 1993

[31] L. Bombelli, J. Lee, D. Meyer, R. Sorkin : "Space-Time as a Causal Set", Phys. Rev. Lett. 59 (1987) 521

[32] R. Sorkin : "A Specimen of Theory Construction from Quantum Gravity" which appeared in "The Creation of Ideas in Physics", ed. J. Leplin, Kluwer, Dordrecht 1995, gr-qc/9511063

[33] M. Frechet : "sur quelques points du calcul fonctionnel", Rend. Circ. Mat. Palermo 22 (1906) 1.

[34] F. Hausdorff : "Grundzuge der mengenlehre" Leipzing, Veit und Comp (1914).

[35] K. Menger : "Statistical Metrics", Proced. Nat. Acad. Sci. USA, 28 (1942) 535.

[36] A. Spacek : "Note on Menger's Probabilistic Geometry", Czechoslovak, Math. Journ. 6 (1956) 72

[37] R. R. Stevens : "Metrically generated probabilistic metric spaces", Fund. Math. 61 (1968) 259

[38] H. Sherwood : "On E-spaces and their relation to other classes of probabilistic metric spaces", J. London, Math. Soc. 44 (1969) 441.

[39] B. Schweizer, A. Sklar : "Statistical Metric Spaces", Pacific J. Math. 10 (1960) 313.

[40] K. Menger : "Geometry and Positivism. a Probabilistic Microgeometry" in K. Menger, Selected Papers in Logic and Foundation p. 225 Ed. H. L. Mulder, Reidel Publ. Comp. Dordrecht 1979

[41] H. Poincaré : "Science and Hypothesis", Dover Pub., N. Y. 1952

Appendix 2. Vibrations and Forms

Ralph Abraham

Based upon a presentation to the third conference on Science and Consciousness, Ramakrishna Mission Institute of Culture, Gol Park, Kolkata, West Bengal (India) on 14 January 2006.

Contents
1. Introduction
2. Personal experiences of vibrations and forms in actual consciousness, 1967-1972
3. My miracle year, 1972 3a. Winter 1972, Paris 3b. Summer and Fall 1972, Nainital
4. The vibration metaphor for levels of consciousness
5. Beyond maps of consciousness: communication between levels
6. Personal experiences of vibrations and forms in artificial consciousness, 1974-1996
7. Conclusion References

1. Introduction

My main goal in this paper is to give an idea, especially a visual idea, of my experiments with vibrations and forms in consciousness, over the past thirty years. The visual representations, computer graphic animations, may be best understood in the context of my personal experiences in actual consciousness exploration during the years 1967 to 1972 which motivated the work, and the philosophical frames, or maps of consciousness, in which I am trying to understand my experiences. These maps are based jointly on my own experiences, and on the philosophies of Greek, Jewish, and Indian origin. I must thank Dr. Paul Lee for his tutelage on the Platonic and Neoplatonic philosophies of the Greek tradition, Dr. Sen Sharma for his explanations of the Kashmiri Shaivite or Trika philosophy and other features of the Indian tradition, and Swami Prabhananda and the Ramakrishna Mission Institute of Culture for extraordinary hospitality during my month in Calcutta, and the privilege of attending this fascinating meeting.

2. Personal experiences of vibrations and forms in actual consciousness, 1967-1972

My story begins in 1967, when I was a professor of mathematics at Princeton University. This is a wonderful university, especially for mathematics, and I was privileged to have colleagues and undergraduate and graduate students, whom I remember fondly to this day. Also, the 1960s was the time of student political unrest, and concomitantly, the time of the Beatles, and the Hip Subculture, or "sex, drugs, and rock and roll", as they used to say. My wonderful students were involved in both of these popular movements, and through them, I also became involved.

In 1967, the three notorious and defrocked psychology professors of Harvard University -Timothy Leary, Richard Alpert (later aka Baba Ram Dass), and Ralph Metzner – were barnstorming about the USA plumping the powers of LSD as an agent of spiritual growth. Leary, under the influence of Vedanta and Gayatri Devi of Los Angeles, used to affect Indian dress, and hold forth on Eastern philosophies. I heard their performance in the Lower East Side of New York City, and decided to try LSD and see for myself. One of my undergraduate students helped me onto the path, and my first experience was an epiphany indeed.

Through this epiphany, I became fascinated with the exploration of consciousness, as we called this path, and continued the work in irregular episodes as I followed my career to the University of California at Santa Cruz in 1968, and subsequently to Amsterdam, to Paris, and to Nainital in the Himalayan foothills. In 1973, I returned to Santa Cruz, and migrated from personal explorations back to academic research on consciousness, chaos theory, and other concerns. My walkabout of five years was over, but was to have a lasting effect on all aspects of my life. I had had hundreds of meditations of the sort practiced in Yoga Nidra, that is, lying prone through the night, in the so-called fourth state of consciousness, and amplified by small doses (eg, 25 mg) of LSD. (Saraswati, 1998) Like Yoga Nidra meditation, the LSD experience provides a trip to the fourth state lasting typically about eight hours, during which sleep is held at bay. These sessions were usually done alone, but sometimes in teams of from two up to a dozen or so others, flying, so we thought, in group formation like a flock of birds. Marijuana use was ubiquitous during this period, but in my experience it made no important contribution to my research, and, generally, I avoided it.

At one time, around 1969, we used large doses of DMT, and this period was crucially important to the whole evolution of my mathematical understanding of consciousness, based on geometry, topology, nonlinear dynamics, and the theory of vibrating waves. For in these experiments, although lasting only a few minutes, the reciprocal processes of vibrations producing forms and forms producing vibrations were clearly perceived in abstract visual fields.

Our perspective during this time and later, was gnostic. That is, we rejected teachers and teachings, and sought to discover cosmology for ourselves. Throughout this period, most of us in the Hip Subculture were apprenticing ourselves to teachers of ancient traditions from East, Mideast, and the West, sharing our experiences, traveling to faraway lands to find teachings, and so on. Teachers travelled through California, and we circled the globe in search of them. Personally I experienced yoga, martial arts (judo and aikido), prehistoric moon rituals, musical meditations, fasting and strict diets (eg, macrobiotics), and Native American ceremonies. This was the background of my interest in vibrations and forms in the field of consciousness.

3. My miracle year, 1972

This final year of my walkabout was blessed with two special learning experiences, one in Paris at the beginning of the year, the other in the Himalayan foothills, in the Summer and Fall.

3a. Winter 1972, Paris

This was the final year of my walkabout, following which I returned to ordinary reality and my post at the University of California at Santa Cruz, an arduous process taking about a year. I began 1972 as a visiting professor at the University of Amsterdam, teaching catastrophe theory. At the same time, I had a visiting position at the Institut des Hautes Etudes Scientifiques (IHES) at Bures-sur-Yvette outside Paris. I used to commute weekly on the train, which I loved. At this time, IHES was newly formed, and had only two permanent professors, David Ruelle and Rene Thom, both of whom were superb. Thom was one of the great mathematicians of the 20th century, and had received the Fields Medal at the International Congress of mathematicians in 1956 for his work in differential topology. I had met him in 1960 in Berkeley, where we began working together on the foundations of catastrophe theory. During 1966, I had written my first books, Foundations of Mechanics, Transversal Mappings and Flows, and Linear and Multilinear Algebra, while Rene had written his foundational work on catastrophe theory, Structural Stability and Morphogenesis, which I arranged to have published by my publisher, Bill Benjamin.

Early in 1972, Rene and I were both stymied in our work and were browsing the borderlines of science looking for clues. I had been reading Kurt Lewin on topological psychology, and on arriving at IHES one day, I asked Rene what he was working on. He pulled a book from his desk and began showing me photo after photo of familiar forms from nature: spiral galaxies, cell mitosis, sand dunes, and so on. These forms, he said, had been photographed in vibrating water. The book was Kymatik, by Hans Jenny, a medical doctor from Dornach, a suburb of Basel, Switzerland. I was thunderstruck to see images from my meditations on the pages of a book, especially in support of the vibration metaphor of the Pythagoreans.

I immediately called Jenny in Dornach, and he agreed to meet me. I took the train to Basel, and was met at the station by Jenny's son-in-law, Christian Stutten, who drove me to Dornach. Along the way I learned that Dornach was the world headquarters of the Anthroposophy movement founded by Rudolf Steiner, the esoteric Christian follower of Madame Blavatsky's Secret Doctrine, around 1900. Jenny was a follower of Steiner, and lived in Dornach along with many other Anthropops. Jenny greeted me in his home, showed me part of his lab, and an animated film of some experiments in progress. I collected his papers and books and went home to Paris and Amsterdam inspired.

As the winter progressed, I thought much about morphogenesis and the mathematics of coupled systems of vibrating membranes and fluids, while continuing to teach catastrophe theory in Amsterdam, and giving many lectures on these subjects at universities all over Europe. Also, my chemically assisted meditations continued, and in them, I pursued the vibration metaphor in conceptual space, and simultaneously, in experiential space.

These experiences were dominated by rapidly vibrating patterns of brightly colored abstract forms, somewhat like the video art and rock concert light shows of the 1960s. The scintillating light caustics projected by the bright sun on the bottom of a swimming pool also give an intimation of the visual aspect of these meditations. An excellent computer simulation has been achieved by Scott Draves in his art works called Electric Sheep, and may be seen on his website. (www.draves.com)

3b. Summer and Fall 1972, Nainital

Suddenly, the spring semester in Amsterdam was over, grades were recorded, and I had a small savings account. It occurred to me to pay India a brief visit before school began again in the Fall of 1972. Here I was influenced by the ambiance of Amsterdam culture, in which I met so many people who had just returned from, or were about to go again to, India. One young man just returned told me how he organized his explorations of the Himalaya: just sit in a tea shop until somebody offers you an experience, then accept it, he said. Just go with the flow. This was my plan. One day at the Kosmos, a psychedelic and meditation hall run by the Dutch government (bless it), I looked up and saw my old friend Baba Ram Dass. The former Richard Alpert, he was among the Harvard trio of professors who had encouraged my decision to experiment with LSD in 1967. Then he had lived briefly in my house in Santa Cruz, California. He had stayed for a time in Nainital, near the western border of Nepal in the Himalayan foothills, where he became attached to a guru called Neem Karoli Baba. I told Baba Ram Dass about my plan to visit India and he gave me instructions for connecting with Neem Karoli Baba. Find your way to Nainital, he said, then hang out at this particular hotel, and if I was supposed to meet Neem Karoli Baba, somebody would approach me and take me to the ashram outside Kainchi, a small village.

And so, late in June, 1972, it came to pass. I went to the ashram with a group of western devotees in a taxi. But on arrival I felt a bit disappointed by the amplified music and carnival atmosphere. I saw the devotees sitting in darshan formation in front of Neem Karoli Baba on his tucket, all in silence. Something seemed to be going on but I was blind to it. Someone would give him prasad, a fruit for example, and he would immediately toss it to someone else. I went back to the hotel in Nainital determined to go on with whomever next approached me.

This process took no time at all. Once back at the hotel, I meet a young barefoot Canadian dressed in a simple smock. He introduced himself as Shambu. As I had been on the road for a long while with a highly evolved travel kit that fit into a small shoulder bag, I was greatly impressed by his kit, which required not even a bag. Shambu explained that he had been living in a cave in the jungle for several months with two other saddhus. There were three small caves by a stream in the jungle, two miles from the nearest town. One of the saddhus had just left, and the village had dispatched Shambu to find a replacement. Apparently the villagers felt their prosperity was only possible with all three caves occupied by appropriate persons engaged in full-time spiritual practice. Smoking ganja apparently counted as spiritual practice, worship of Shiva it seems. Shambu was sure that he had been guided to me as I was the chosen person. Shambu put me on a bus with the usual sort of instruction: ride the bus to the end of the line at Almora, from there I would be guided somehow. This was monsoon season, and there had been heavy rain. After a short while the bus was firmly halted by a major road washout. Everyone climbed out of the bus. Looking down the slope, I was surprised to see Neem Karoli Baba's ashram for the second time. What a coincidence! Then someone came out to say I should come in at once, as Neem Karoli Baba was asking for me. Was this really happening, or was there some mistake? Neem Karoli Baba gave me a bag of breakfast cereal. He said I was going to need it in the jungle. Two young Indian devotees were told to guide me on a trek through the jungle around the washout, and put me on a bus for Almora on the other side. By this time I was losing my Western mind, and all this seemed more like paranormal phenomena than conspiracy theory.

It was midnight when finally the second bus arrived in Almora. The village was dark, but moonlight through a clearing in the clouds showed the shops in silhouette. A man descended from the bus after me. He had a bearer with a long box balanced on his head. I asked him where he was going, hoping for a clue for my next steps. He said that he was a student of Jim Corbett, the famous hunter of man eating tigers. I had just read Corbett's book, Maneaters of the Kumoan. Actually, we were now in the Kumoan Hills. The man said the long package was his rifle. There was a maneating panther on the loose nearby, and he was about to spend the night in a tree overlooking a fresh human kill, hoping to shoot the panther. This was his job, he had been sent by the government. I decided not to follow him into the jungle.

I followed some other people who descended from the bus. They seemed to know where they were going, on a footpath into the jungle. One by one they vanished into side paths, and then I was walking alone into the dark unknown, following this single-track footpath. I could not stop to sleep, for fear of the panther. As long as the path continued, and looked like it was used by humans, I would continue, until I found where it went. Another village or whatever. Seemed like a plan, for an hour or so, until there was a fork in the path. In the dark I could see no indication which way to go. Just then I was startled by a rustle very close by. I could see only grey on grey in the darkness. Then a voice said in clear English, "Good evening saheb, I am from the Wisdom Garden School. I have been waiting for you. You are to go this way". Then he pointed to the left fork, and vanished. So on I went, until I heard voices. Following the sound, I came upon a group of Western hippies in a house, who offered me a place to sleep. Apparently this was the Kasa Devi Ridge, where the German Lama Govinda had established himself some years ago, after going totally native in the Himalaya. In the morning they showed me the way to a village nearby, which was Dinapani,

Abraham and Roy

my destination. The headman interviewed me in his chai shop, approved me for cave service, and asked his young son to guide me into the jungle to the cave.

Indeed there were three caves and two jungle babas, who were muni, that is, they did not speak. Not out loud at least. But voices in my head made me welcome, and spelled out the rules. I must keep a fire going in my cave every night, or a panther would come to claim the space. I must go to the stream every morning to wash, and worship Shiva in an underwater grotto that has been used for centuries and has a polished lingam. The dhuni (small ritual fire) must be kept going. Food would be brought by villagers every morning on their way into the forest to tap turpentine trees.

All went well for a week or so. I thought of writing my mother to say I had found a place where I should stay for a few months to further my education, but I could not manage to write. Every night I practiced my yoga nidra, and explored further the vibrational realms. There seemed to be instruction regarding the use of 'tools of light" for self-defense and self-maintenance. I practiced, according to these instructions, during the day, while sitting meditation by the dhuni after my bath with Shiva and the daily meal of dhalbhat (rice and lentils), gor (raw sugar), and the mandatory chillum (straight pipe) of hashish.

Then the trouble began. I had some unwelcome orders during the night. I was to leave this place immediately. I resisted. Then the orders were repeated with physical discomforts, which would go away as soon as I agreed to leave in the morning. But in the morning I changed my mind. And so on, in a cycle.

Until one day, around my 36th birthday, July 4, while the other two yogis were away on mysterious missions and I was hard at work meditating by the dhuni, I saw a person approaching, far down the jungle path. This figure got larger and larger, and eventually resolved into a vision from hell, a wild man with a spear, clothed primarily in ashes. He sat down by the fire and accepted a toke from my fully loaded chillum. My paranoia subsided, as apparently he meant no harm. After an hour or so staring into the distance, he turned to me and spoke in unaccented American, "Don't you understand, you are supposed to leave here. I am going to get up and leave now, and you are to follow me". Which he did. And I did, after collecting my small bag from the cave. After a walk of a mile or so down a path I had not seen before, he said, "I am going this way, you go that way", and disappeared around a bend. I followed the indicated jungle path, I am not sure how far, and it led directly to Neem Karoli Baba's ashram. Again, the old fellow was apparently expecting me, bellowing, "Where is that professor from California? Bring him here." And so, reluctantly, began my relationship with Neem Karoli Baba. I was setup with a house, a library of Sanskrit classics in English translation, and a few devotees for company – including one with Sanskrit skills, Kedarnath, his partner, Uma, and their baby, Ganesh, born during one of our meditations. I was informed by Neem Karoli Baba that I had a mission to relate my meditation experiences to the Sanskrit classics, and transmit the understanding somehow to my colleagues in the USA. These sources included the Vedas, a few Upanishads, works by Sri Auribindo, and the Yoga Vasishta, a primary text for the Trika philosophy of Kashmiri Shaivism.

I became known at Veda Vyaasa. I remained in this setup for six months, most of the time with Ray Gwyn Smith, now my wife, who had arrived from California in the meanwhile. The night meditations amplified by microdoses of LSD continued, as I had brought a supply with me from Holland right from the start. Yoga Vasishta was a great inspiration and support for my ideas of vibrations and maps of consciousness. For example:

> VASISHTA replied: There does exist, O Rama, the power or energy of the infinite consciousness, which is in motion all the time; that alone is the reality of all inevitable futuristic events. for it pen-

etrates all the epochs in time. It is by that power that the nature of every object in the universe is ordained. That power (cit sakti) is also known as Mahasatta (the great existence), Mahaciti (the great intelligence), Mahasakti (the great power), Mahadrsti (the great vision), Mahakriya (the great doer or doing), Mahadbhava (the great becoming), Mahaspanda (the great vibration). It is this power that endows everything with its characteristic quality. (Venkatesananda, 1993; p. 89)

Neem Karoli Baba and the entire satsang departed for warmer climes to the south, after the thermometer in Nainital dropped below freezing in October. Ray and I departed in December for a Himalayan trek in Nepal, where I donated my library to a local university. We walked about 400 miles and returned to California early in 1973. And thus ended my miracle year,1972, and also the five year period of one-point focus on spiritual exploration. After returning to Santa Cruz and my job as math professor at UCSC, I reinterpreted the mission given me by Neem Karoli Baba as a program of academic research on vibrations and forms in mathematical models, and in physical fluids as well.

What I learned about cosmos and consciousness during this final year of the five-year project cannot be said in words, perhaps mathematics will be helpful. I imagined this as my task intended by Neem Karoli Baba. But I had to go on alone, as both Neem Karoli Baba, and Hans Jenny died at this time.

4. The vibration metaphor for levels of consciousness

All my experiences in inner research conformed to the conceptual framework of levels. These levels of consciousness are alternate realities, that may be experienced only one at a time. In the meditation experience, they are transited, in a sequence, from ordinary reality to more abstract levels. This framework is well known from the Greek, Jewish, and Indian traditions, as we describe below.

4a. My experience

The same levels of consciousness always appeared in the same order of increasing abstraction, and were recognizable as forms of reality. They seemed as real as ordinary reality. With successive visits, they always had the same recognizable characteristics: visual aspects, colors, speeds of vibration, typical forms. I thought of these levels, each having its own spatial and temporal dimensions, as being stacked up in another dimension, like horizontal planes stacked vertically, with the more abstract levels "higher". In fact, we spoke of these meditations as "getting high". We spoke of the ascent to higher consciousness. At the end of a meditation, we would descend through these levels in the order reverse to the ascent. This was commonly called "coming down". The whole meditation was called a 'trip", like a stairway to heaven and back.

The lowest level, ordinary reality, as we all experience it in everyday perception, is matter-like. It is the world of matter and energy, spatially localized, ego centered, and so on. Things are objects. Philosophers may speak of vibrations or vital forces, but we do not normally observe them.

On the next level, ordinary objects appear the ordinary way, but are seen to have "vibratory fields" or "auras" around them, They are surrounded for a short distance by these shimmering auras. In this level or reality or consciousness, we may interact with objects in the ordinary way, for example by touching, and observe the ordinary response, and also a reaction of the objects' auras. What do I mean by "vibratory fields"? This can be best answered by computer graphic animations that simulate my visions quite well, and that is one reason for my research with analog and digital simulations of artificial consciousness over these past thirty years, which I am going to discuss below. Meanwhile, you might just think of the patterns of light caustics on the bottom of a swimming pool, from the bright sun overhead, as you paddle about on the surface of the water, looking down. That kind of moving image, in brilliant colors, changing with great rapidity – all the time appearing meaningful in a mysterious way, as the abstract visual music seems familiar as deja vu – is exemplary of my idea of a vibration: visual music in air, light through water, waves on the ocean, and so on. The "field" is the unknown medium that supports the vibration in consciousness, as water supports the waves on the ocean. (Hesse, 1961) In the next level up, the object aspect is greatly reduced, and auras predominate. And higher yet, objects vanish, and the auras join together into a single cosmic vibrating field. Parts of the field seem to behave like objects or beings or disembodied entities. It is possible to navigate and move about the field in some sense, or rather, to move the focus of attention by an exertion of will. Attention replaces the self, in that the self seems everywhere, but attention can be stopped-down, focused, panned, and zoomed-in, as it were. One is everywhere but there is still a personal center of awareness. Yet above, there is nothing but the field, and that is as far as I have gone. but I do not think that the "pure consciousness" experience of no thought is the end of the line.

This is the essence of my recollection of these indelible experiences of long ago, up to 1972. I have maintained them to some extent by less extreme forms of mediation over the years, but much detail has been lost. If my description sounds like every other description of mystical experience, that is most likely due to a universality of the experience. I always had the conviction that the experience is universal, but the transcriptions into words vary.

After my return to academia in 1974, a decade was to pass in mathematical research and teaching, before I could resume my study of the philosophical and cosmological traditions that might shed light on my experiences from 1967 to 1972. My first focus was the Ancient Greek and Western Esoteric Traditions. Later I turned to early Jewish mysticism, and more recently, I resumed my search of the Indian literature.

4b. The Greek tradition

The maps of consciousness from Ancient Greece have various levels, beginning with the ideas and forms of Pythagoras (570-500 BC), and formalized as a stack of levels by Plato (429-347 BC). Around 360 BC we find four levels described in Plato's Republic. From the top down, these are: Forms, Intellect, Nature, and Shadows. (Shear, 1990; p. 12) Later authors usually refer to the four Platonic levels, described in the later dialogues, as the Good, the Intellect (nous), the Soul (psyche), and Nature (physis). The lore of the soul was extended in the Chaldean Oracles of Julianus (ca 200 AD). (Julianus, 1989) (Lewy, 1956) The Greek map further evolved then in the Neoplatonic sources from Plotinus (205-270 AD) to Ficino (14331499). (Abraham, MS#116, 2006)

With Plotinus, Porphyry (232-304), and Iamblichus (250-326), we have the addition to Plato's scheme of the Spirit (pneuma). Also known as the Vehicle of the Soul (okema), this was part of the Neoplatonic theory of incarnation of the individual soul, in which a soul descended through layers of increasing density, being wrapped in Spirit (emanated from stars and planets) en route to incarnation and birth. (Walker, 2000; p. 38) The Spirit mediated between the incorporeal soul and the corporeal body, and supported the functions of sense perception and imagination. (Finamore, 1985; p. 1-2) Later, especially in the theology of Proclus (409-487) and Ficino, the Spirit provided the basis of astrological influence: the ongoing astrological contact between the soul and the planets. (Moore, 1982; p. 53) This theory of astrological influence survived in the works of Kepler (1571-1630). (Rabin, 1987; ch. 3) (Kepler, 1997; bk. 4)

Relating all this to my direct experience, I identified my matrix (intermediating vibrationally between all adjacent levels, described below) with the Neoplatonic Spirit. But as far as the vibration metaphor is concerned, we have from the Greek tradition, as far as I know, only the harmony of the spheres concept from Pythagoras, Ptolemy, and Kepler. These sources offer abstract concepts, but there is no record of experiences obtained by meditation. Also, the harmony envisioned in the Greek tradition is only that of harmonious sounding dyads (pairs of musical tones), and not the vibration/form duality of my experience. For this, we know of no antecedent before Ernst Chladni (1756-1827), who founded acoustic physics around 1800, and inspired Hans Jenny.

This is a subtle yet important distinction: the vibration of Pythagoras, Ptolemy, and Kepler is one-dimensional, the musical vibration of a plucked string. Harmony for them is the musical consonance of two plucked strings, the tones related by the ratio of the lengths of the two strings. But the vibration of Chladni, Jenny, and me is two-dimensional, the musical vibration of a stuck flexible membrane or plate. Harmony for us is a matter of the forms created by a vibration of dimension two or more – as the forms seen in meditation, in the higher levels of consciousness, for example.

4c. The Jewish tradition

Early Jewish esotericism and mysticism derived from Philo Judeus, Greek Gnosticsm, and Eastern sources in the early Christian era, especially in Alexandria. (Scholem, 1978: pp. 8-21) The Merkabah tradition, it seems to me, is a coded story of early explorers of my own path. These pioneers would go down to the basement to spend the night in meditations guided by concentration on visual images, and amplified by breathing exercises. The path desired was an ascent through seven levels of increasing abstraction, each identified by visual features of abstract animations. (Blumenthal, 1978; ch. 5)

4d. The Indian tradition

The Indian tradition provides a number of different schemes for levels of consciousness, including the five koshas, seven chakras, 36 tattvas, and so on. The closest scheme to that of my own experience is that of the five koshas. These are, from the top down: the bliss body (anandamaya kosha), astral body (vijnanamaya kosha), mental body (manomaya kosha) pranic body (pranamaya kosha), and the food body (annamaya kosha). These subtle bodies, or levels, may be ascended by prolonged practice of yoga nidra, or other meditations, ultimately reaching the bliss body. The bliss body is described as an experience of total transcendence, where only the fundamental vibration of the unconscious system remains. (Saraswati, 1998; p. 54)

The vibration metaphor that I encountered in the Yoga Vasishta explicitly entered the Indian literature in the Spanda (vibration), Urmi (wave), and Prana (life-force) concepts of Trika philosophy (Kashmiri Shaivism) due to Vasugupta, his disciple, Kallata, and his student in turn, Abhinava Gupta, tenth century AD. I am a beginning student of this tradition, and I am grateful to Prof. D. Sen Sharma of Calcutta for leading me to this historical information. (Sen Sharma, 2003, 2004; Dyczkowski, 1992; Singh, 1980.)

Pythagoras may have visited India. And it is known that there were yogis in Ancient Greece; they were called gymnosophists.

So vibration metaphors might have diffused either way. The origin of hte vehicle of the soul has been traced to Babylonia. (Lewy, 1978; p. 413)

5. Beyond maps of consciousness: communication between levels

In Section 4a above I have set out the cosmographic (map of consciousness) that I had obtained before 1972, with the levels of consciousness stacked up, with ordinary reality and the individual soul or microcosmic levels at the bottom, the cosmic or macrocosmic levels above, and the mesocosmic levels interpolating in between. This personal cartography, although supported by received literature of all traditions, was lacking any model for the interaction or communication between levels. It was in 1972, especially in the cave near Almora, that this part of the picture was filled in. I can express this best in the mathematics of chaos theory, but here I will try in words. First of all, we see in Hans Jenny's books – and in my continuation of his work in my fluid dynamical vibration laboratory at UCSC in the years 1974 to 1980 – how a vibration creates a form. Similarly, a form impressed upon a spontaneously vibrating field modifies that vibration field, and results in a new vibration that encodes the form. Vibrations to forms, forms to vibrations, somewhat like the particle-wave duality of modern atomic physics. Okay, lets use this idea to connect levels of consciousness.

Consider just two of the levels, that are adjacent in the traditional cosmographic map described above, and each in a state of vibration, as we experience them separately in our meditations. In meditation, we experience a sort of quantum leap ascending, and also descending, between levels. We cannot directly perceive any connection or semaphoric transmissions in the space between levels. For this we are grateful for suggestions from the received literature of the rishis of East and West, who show us how to observe these hidden communications. The suggestions I found useful in 1972 were found in Yoga Vasishta, and I am grateful to Neem Karoli Baba for that. Subsequently, I discovered the Spanda literature of the Trika or Kashmiri Shaivite philosophical tradition, thanks to Professor D. Sen Sharma of the Research Department of the Ramakrishna Mission Institute of Culture, Kolkata.

This is the idea. In the space between levels there is yet another, finer, vibratory field, that I will call the matrix. This resembles the diffusion of neurotransmitters in the extracellular space between neurons in the mammalian brain. The vibration in level A creates a form in level A, this is impressed in the intermediate matrix field, modulating the ongoing vibration there, which carries a vibratory signal to level B, where it impresses a form on level B, and that creates a vibration in level B. This semiotic process, mediated by the matrix field, is hard to grasp in words. However, I have created (with help of Peter Broadwell) a computer graphic simulation, which is easily grasped. But this was not possible until the 1990s. (Abraham, MS#86, 86B)

6. Personal experiences of vibrations and forms in artificial consciousness, 1974-1996

My experiments of vibrations and forms in actual consciousness of 1972 morphed, after my return to academic life in 1974, into a program of laboratory science modeled on the work of Hans Jenny. I like to think of this as research in artificial consciousness, but of course it was more practical to call it fluid dynamics.

The apparatus – I called it a macroscope – consisted of a coupled system of vibrations in various levels. At bottom was an electronic oscillator capable of producing sine waves, square waves, saw tooth waves, and so on, with control knobs for frequency and amplitude. This source was converted into up-down mechanical vibrations by a horizontal high-fidelity loudspeaker, that in turn vibrated a column of air above the speaker cone. And this moved a transparent membrane, and above that, a thin layer of water, in which was activated a pattern of thin water waves. These waves were imaged on a translucent screen by an optical system containing two telescope mirrors and a point source of light.

These layers could be regarded as a crude model for levels of consciousness, in that the vibrations of one level created forms on another level, and vice versa. Video recordings of the moving patterns of light on the screen were very reminiscent of the visual experiences seen in my meditations. Some of this work was reported during the 1970s. (Abraham, MS#14-20) After 1975 or so, mathematical models and computer simulations gradually replaced the analog simulations with the macroscope, and computer graphic video recordings have provided some moving patterns that are highly suggestive of the visual component of my meditation experiences. (Abraham, MS#25-86)

In particular, the papers MS#86 and 86B describe a simulation of an experiment by Rupert Sheldrake on telepathy from a person to her dog. In our model, a vibrating two-dimensional field was modulated by the introduction of a geometrical form representing the person's thought to come home, and this modulation was recorded as a memory engram of the patterns perceived over time on a one-dimensional "retina" in the dog's mind. This clearly shows the role of memory in consciousness, as described by Henri Bergson and in Kashmiri Shaivism. (Chakrabarti, 2004)

7. Conclusion

A personal odyssey through the spiritual practices of several traditions, begun in 1967 and still ongoing, has motivated a research program in chaos theory and computational mathematics. The products of this mathematical program, unlike the subjective experiences of meditation, are open to the scientific paradigm of publication, replication, and the hermeneutical circle of theoretical and experimental synergy. This program belongs to the category of mathematics of consciousness begun by Pythagoras, Plato, and the Sanskrit classics, rather than that of science and consciousness, but may have some implications for science in the long run.

REFERENCES

Papers by Ralph Abraham

Papers numbered above MS#67 may be viewed on the web at: http://www.ralph-abraham.org/articles.

MS#14. Psychotronic vibrations. First Int'l Congress Psychotronics and Parapsychology, Prague, 1973.

MS#15. Vibrations and the realization of form. In: *Evolution in the Human World*, (Jantsch and Waddington, eds.) Addison-Wesley, Reading, 1976, pp. 134-149.

MS#16. The macroscopy of resonance. In: *Structural Stability, The Theory of Catastrophes, and Applications*, Springer, New York (Lecture Notes in Mathematics, Vol. 525), 1976, pp. 1-9.

MS#17. Simulation of cascades by videofeedback. In: *Structural Stability, The Theory of Catastrophes, and Applications*, Springer, New York (Lecture Notes in Mathematics, Vol. 525), 1976, pp. 10-14.

MS#18. Dynasim: exploratory research in bifurcations using interactive computer graphics. *Ann. N.Y. Acad. Sciences* 316

(1976), 673-684.

MS#20. The function of mathematics in the evolution of the noosphere. In: *The Evolutionary Vision*, (E. Jantsch, ed.), AAAS Selected Symposium Ser., 1981, pp. 153-168.

MS#25. Dynamical models for thought, *J. Social Biol. Structures*, 1985, pp. 13-26.

MS#42. Vibrations in math, music and mysticism, *IS Journal* 1, 1(0) 7-8 (1986).

MS#44. Mechanics of resonance, *Revision*, 10(2): 13-19 (1987).

MS#47. Visual musical instruments, *High Frontiers*, Fall, 1988.

MS#52. Order and chaos in the toral logistic lattice (with John Corliss and John Dorband), *Int. J. Bifurcation and Chaos*, 1(1):227-234 (March, 1991).

MS#58. Visualization techniques for cellular dynamata, in: *Introduction to Nonlinear Physics*, Lui Lam, ed., Springer-Verlag, 1990.

MS#60. Erodynamics and cognitive maps, In: *New Paradigms for the 21st Century: The Evolution of Contemporary Cognitive Maps*, Ervin Laszlo and Ignazio Masulli, eds., Gordon and Breach, 1991.

MS#73. Human fractals, the arabesque in our mind. *Visual Anthropology Review*, 9, 1993, pp. 52-55.

MS#86. Vibrations: communication through a morphic field, *Proc. Intl. Conf. Synthesis of Science and Religion*, Calcutta, 1996.

MS#86B. Vibrations: communication through a morphic field,

Pt 2. (with Peter Broadwell). Preprint.

MS#105. Vibrational resonance and cognitive internalization. In: *Proceedings of Einstein Days in Visva Bharati University, Santiniketan, West Bengal, India, March 15-18, 2000: International Seminar on Cognitive Processes of Internalization in Humanities and Sciences.*

MS#106. A two worlds model for consciousness. Preprint.

MS#116. The death and rebirth of the world soul, 2500 BCE – 2005 CE, a concise overview. In: Ervin Laszlo, *Science and the Reenchantment of the Cosmos: The Rise of the Integral Vision of Reality*, Rochester, VT: Inner Traditions, 2005.

Books

Blumenthal, David R. *Understanding Jewish Mysticism: A Source Reader. The Merkabah Tradition and the Zoharic Tradition.* New York: Ktav, 1978.

Dyczkowski, Mark S. G. *The Stanzas on Vibration.* Albany, NY: SUNY Press, 1992.

Finamore, John F. *Iamblichus and the Theory of the Vehicle of the Soul.* Chico, CA: Scholars Press: 1985.

Hesse, Mary B. *Force and Fields: the Concept of Action at a Distance in the History of Physics.* London: Nelson, 1961.

Julianus. *The Chaldean Oracles*, transl. T. Stanley. Berkeley Heights, NJ: Heptangle Books: 1989.

Kepler, Johannes. *The Harmony of the World.* Transl. A. J. Aiton, A. M. Duncan, and J. V. Field. New York: American

Philosophical Society, 1997.

Lewy, Hans. *Chaldean Oracles and Theurgy: Mysticism, Magic, and Platonism in the Later Roman Empire.* Cairo: l'Institute Francais d'Archeologie Orientale, 1956.

Moore, Thomas. *The Planets Within: Marsilio Ficino's Astrological Psychology.* Lewisburg: Bucknell Univ. Press: 1982.

Prabhananda, Swami. *Philosophy and Science: An Exploratory Approach to Consciousness.* Kolkata, India: Ramakrishna Mission Institute of Culture, 2003.

Prabhananda, Swami. *Life, Mind, and Consciousness.* Kolkata, India: Ramakrishna Mission Institute of Culture, 2004.

Rabin, Sheila J. *wo Renaissance Views of Astrology: Pico and Kepler.* Ph.D. thesis, City University of New York, 1987.

Saraswati, Swami Satyananda. *Yoga Nidra.* Munger, Bihar, India: Yoga Publications, 1998.

Scholem, Gershom. *Kabbalah.* New York: New American Library, 1974/1978.

Shear, Jonathan. *The Inner Dimension: Philosophy and the Experience of Consciousness.* New York: Peter Lang, 1990.

Singh, Jaideva. *Spanda-karikas: The Divine Creative Pulsation.* Delhi: Motilal Benarsidass, 1980.

Venkatesananda, Swami. *Vasishta's Yoga.* Albany, NY: SUNY Press, 1989.

Walker, D. P. *Spiritual and Demonic Magic from Ficino to Campanella.* University Park, PA: Pennsylvania Univ. Press: 1958/2000.

Articles

Chakrabarti, Arindam. Matter, memory, and unity of the self. In: Prabhananda, 2004; p. 67. Roy, Sisir. Quantum information and levels of consciousness. In: Prabhananda, 2003; pp. 223-241.

Sen Sharma, D. Consciousness in Indian philosophical thought with special reference to the Advaita Saiva School of Kashmir. In: Prabhananda, 2003; pp. 207-222.

Sen Sharma, D. Concept of prana in Kashmir Saivism. In: Prabhananda, 2004; pp. 515-517.

Websites

Draves, Scott. http://www.draves.org.

Neem Karoli Baba Ashram. http://www.nkbashram.org

FIGURES

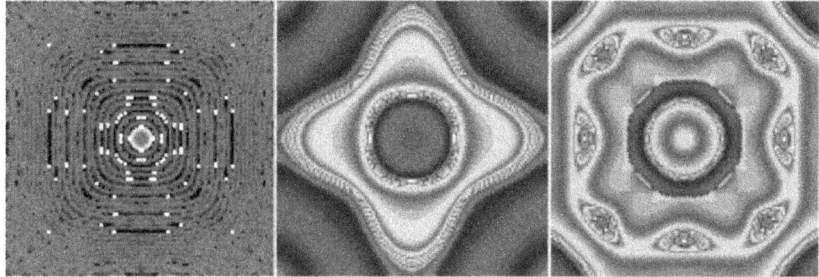

Figure 1. Successive stages of morphogenesis in a lattice of oscillators: the Brusselator simulation of a chemical vibration. From joint work with John B. Corliss, Central European University, Budapest, Hungary.

Figure 2. Vibration in an experiment of artificial consciousness: simulation of the Sheldrake experiment. From joint work with Peter Broadwell, Palo Alto, California, USA.

Appendix 3.
Technical Summary of the AR Process

The AR process is not a description of physical reality, but just a mathematical model that captures some aspects of our experience of physical reality. We will summarize this process in three stages, using mathematical notations.

- A. We begin with a description of our microscopic system, QX.

- B. Then we will go on to extract from it our macroscopic system, ST.

- C. Finally, we describe in summary the embedding of ST into an Euclidean space, EST.

A1. There is a finite, but huge, point set, which is static throughout the process. Let S denote this finite set. Enumerate this set by fixing a bijection from S to N, the cardinality of S. Thus, S is a set of points, $\{n_0, n_1, ..., n_{(N-1)}\}$. These points are called *nodes*.

A2. At each node and each moment of time there is an internal node-state, which is some number of quanta of information. Thus, we have a set of time-dependent node-states, $\{s_0, s_1, ..., s_{N-1}\}$.

A3. There are no bilateral connections. That is, for each pair of nodes, n_i and n_j, there may be a directed link from n_i to n_j, or none. We agree there cannot be a directed link from n_i to n_j if there is one from n_j to n_i.

A4. There is a global time clock for the system. The time variable, t, is a natural number, and increases by one at regular intervals, called *clicks*.

A5. The directed links may appear, disappear, or change direction, with each click. They change according to a fixed dynamical rule.

A6. With each click, each node n_i sends one quantum of information to the node n_j if there is a directed link from n_i to n_j.

A7. At each time there is digraph, a directed graph on S, defined by the directed links. Let $D(t)$ denote the state of this digraph at clock time t, an integer. Associated with $D(t)$ is a graph $G(t)$, in which the directions of $D(t)$ are ignored.

This is our microscopic system, QX. Next we will describe the emergence of the macroscopic ST system from QX, that is, the process $QX \to ST$.

B1. For each node, n_i, of $D(t)$ let w_i denote its node-weight, that is, the number of directed links of $D(t)$ that either arrive at, or depart from, n_i. Thus, we have a finite sequence of node-weights, $(w_0, w_1, ..., w_{(N-1)})$.

B2. Next, at each time, t, we may construct, from the digraph $D(t)$, a permutation of the set S of nodes, as follows. We reorder the nodes of S according to their node-weights, in decreasing order. If several nodes have the same node-weight, we retain their original order. Let $P(t)$ denote the permutation of N obtained in this way.

B3. A clique of a permutation is a maximal inverse sequence. Compute the cliques of $P(t)$. This may be done by inspection if N is not too large. Let $K(t)$ denote the set of all cliques of $P(t)$. These cliques, which are simply subsets of $\{0, 1, ..., n-1\}$ in decreasing order, will be considered the supernodes of our macroscopic system, ST.

B4. If K is a finite set of natural numbers, let the span of K denote the filled-in interval, $span(K) = [min(K), max(K)]$. We define a superbond between two supernodes, or cliques, if and only if their spans are disjoint. Thus we have a graph $ST(t)$ defined by these supernodes and superbonds.

This is our macroscopic system, ST. Finally we will describe the pseudo-isometric embedding of ST into a Euclidean space, $ST \to EST$, following Chapter 7.

C1. For every pair of disjoint cliques of $K(t)$, we define their overlap, a measure of the entanglement of the two cliques, by counting points in the intersection and union of the sets spanned by the two cliques. Details and examples may be found in Chapter 8. These overlap measurements may be used to define distances: more entanglement corresponding to a smaller distance.

C2. Embed $K(t)$ in a Euclidean space, and relax the embedding to approximate as closely as possible an isometry. That is, the distance between the images of two cliques represents their entanglement.

The process $QX \to ST \to EST$ is called *condensation*.

References

- Abraham, Ralph (1994). *Chaos, Gaia, Eros: A Chaos Pioneer Uncovers the Three Great Streams of History.* San Francisco: Harper Collins.

- Abraham, Ralph (MS#105, 2000). Vibrational Resonance and Cognitive Internalization. Preprint.

- Abraham, Ralph (MS#116, 2006). An overview of the death and rebirth of the world soul, 2500 BCE – 2005 CE, a concise overview. In (Laszlo, 2006; pp. 177-186).

- Abraham, Ralph (MS#118, 2006), Vibrations and Forms. In: *Consciousness: A Deeper Scientific Search*, J. Shear and S. P. Mukherjee, eds. Kolkata, India: Ramakrishna Mission Institute of Culture, pp. 192-212.

- Abraham, Ralph (MS#128, 2008). Consciousness and the New Math, Preprint.

- Abraham Ralph and Roy Sisir (MS#119, 2007). The Planck Scale and agent based simulations of quantum space-time, *Int.J. Pure and Applied Mathematics*, 39:4 (2007) pp. 445-458.

- Abraham, Ralph, and Sisir Roy (MS#122, 2007). A Digital Solution to the Mind/Body Problem. Preprint.

- Aitchison, I. J. R. (2009). Nothing's Plenty. The Vacuum in Modern Quantum Field Theory. *Contemporary Physics, 50:1,261–319.*

- Aurobindo, Sri (1996) *The Synthesis of Yoga.* Twin Lakes, WI: Lotus.

- Aurobindo, Sri (1997) *The Integral Yoga.* Twin Lakes, WI: Lotus.

- Balslev, Anindita Niyogi (2009). *A Study of Time in Indian Philosophy, Third Edition.* Delhi: Motilal Banarasidass.

- Barad, Karen (2007). *Meeting the Universe Halfway: Quantum Physics and the Entanglement of Matter and Meaning* Durham, NC: Duke University Press.

- Barbour, Julian (1999). *The End of Time: The Next Revolution in Physics.* Oxford: Oxford University Press.

- Bell, John L. (2005). The Continuous and the Discrete in Ancient Greece, the Orient, and the European Middle Ages. Chapter 1 in: *The Continuous and the Infinitesimal in Mathematics and Philosophy.* Monza: Polimetrica.

- Bentov, Itzhak (1977). *Stalking the Wild Pendulum: On the Mechanics of Consciousness* New York, NY: Dutton.

- Bernstein, Jeremy (1991). *Quantum Profiles.* Princeton, NJ: Princeton University Press.

- Boi, Luciano. (2009). Creating the physical world ex nihilo? On the quantum vacuum and its fluctuations. Preprint.

- Braden, Gregg (1996). *Awakening to Zero Point: The Collective Initiation* Bellevue, WA: Radio Bookstore Press,

- Braden, Gregg (2007). *The Divine Matrix: Bridging Space, Time, Miracles, and Belief* Carlsbad,Ca: Hay House.

- Broughton, Richard (1991). *Parapsychology: A Controversial Science.* New York : Ballantine Books.

- Burtt, Edwin Arthur (1927/1959). *The Metaphysical Foundations of Modern Physical Science: a Historical and Critical Essay.* London: Routledge and K. Paul.

- Cahill, Reginald T. (2008). Process physics and Whitehead: The new science of space and time. In: F. Riffert and H.-J. Sander (Eds.), *Researching with Whitehead: System and Adventure (pp. 83-126)*. Munich: Verlag Karl Alber.

- Cahill, Reginald T. (2005). *Process Physics: From Information Theory to Quantum Space and Matter*. New York: Nova Science.

- Capra, Fritjof (1975). *The Tao of Physics: An Exploration of the Parallels between Modern Physics and Eastern Mysticism*. Boston: Shambala.

- Combs, Allan (2009). *Consciousness Explained Better: Towards an Integral Understanding of the Multifaceted Nature of Consciousness*. St. Paul, MN: Paragon.

- Corbin, Henri (1977). *Spiritual Body and Celestial Earth: From Mazdean Iran to Shī'ite Iran*. Princeton, NJ: Princeton University Press.

- Daitz, Stephen G. ((1981/1984). *The Pronunciation and Reading of Ancient Greek*. New York: Jeffrey Norton.

- Dasgupta, S .N. (1974). *Yoga Philosophy*. Delhi: Motilal Banarsidass.

- Davies, P.C.W., and J.R. Brown (1986/1999). *The Ghost in the Atom: A Discussion of the Mysteries of Quantum Physics*. Cambridge: Cambridge University Press.

- Deleuze, Gilles (1968/1994). *Difference and Repetition*. New York, NY: Columbia University Press.

- Deleuze, Gilles (1988/1993). *The Fold: Leibniz and the Baroque*. Minneapolis, MN: University of Minnesota Press.

- De Quincey, Christian (2002/2010). *Radical Nature: The Soul of Matter*. Rochester, VT: Park Street.

- Dirac, P. A. M. (1930/1935/1947/1958). *The Principles of Quantum Mechanics.* Oxford: Oxford University Press.

- Dyczkowski, Mark S. G. (1987). *The Doctrine on Vibration: an analysis of the doctrines and practices of Kashmir Shaivism.* Albany, NY: SUNY Press.

- Dyczkowski, Mark S. G. (1992). *The Stanzas of Vibration: The Spandakarikas with four commentaries.* Albany, NY: SUNY Press.

- Dyczkowski, Mark S. G. (1992). *The Aphorisms of Siva: the SivaSutra with Bhaskaras commentary, the Virttika.* Albany, NY: SUNY Press.

- Fredkin Edward (2000), www.digitalphilosophy.org.

- Gaukroger, Stephen (1995), *Descartes: An Intellectual Biography.* Oxford: Clarendon Press.

- Geldard, Richard G. (2007). *Parmenides and the Way of Truth.* Rhinebeck, NY: Monkfish.

- Gerald, Curtis F. (1970). *Applied Numerical Analysis.* Reading, MA: Addison-Wesley.

- Goswami, Amit (1993, 1995). *The Self-Aware Universe: How Consciousness Creates the Material World.* New York: Penguin.

- Greenstein, George, and Arthur G. Zajonc (1997/2006). *The Quantum Challenge: Modern Research on the Foundations of Quantum Mechanics*, 2nd edn. Sudbury, MA: Jones and Bartlett.

- Grimes, John (1996). *A Concise Dictionary of Indian Philosophy: Sanskrit Terms Defined in English.* Albany NY: SUNY Press.

- Guenther, Herbert V. (1989). *From Reductionism to Creativity: rDzogs-chen and the New Sciences of Mind.* Boston: Shambala.

- Hagan, Martin T., Howard B. Demuth, and Mark Beale (1996). *Neural Network Design.* PWS Publishing.

- Hahm, David E (1977). *The Origins of Stoic Cosmology.* Columbus, OH: Ohio State Univ. Press.

- Hameroff, Stuart (1987). *Ultimate Computing.* Amsterdam: Elsevier.

- Herbert, Nick (1985). *Quantum Reality: Beyond the New Physics.* New York: Penguin.

- Herbert, Nick (1988). *Faster than Light: Superluminal Loopholes in Physics.* New York: Penguin.

- Herbert, Nick (1993). *Elemental Mind: Human Consciousness and the New Physics.* New York: Penguin.

- Hey, Anthony J. G. ed. (1999). *Feynman and Computation: Exploring the Limits of Computers.* Reading, MA: Perseus Books.

- Hillman, James (1996). *The Soul's Code: In Search of Character and Calling.* New York: Warner Books.

- Husserl, Edmund (1954/1970). *The Crisis of European Sciences and Transcendental Phenomenology.* Evanston, IL: Northwestern University Press.

- Ibn al-'Arabi, 'Alī b. Muhammad (1980). *The Bezels of Wisdom.* Transl. and Introduction by R. W. J. Austin. Mulwah, NJ: Paulist Press.

- Isham C.J. (1995). *Structural issues in quantum gravity.* http://xxx.lanl.gov/gr-qc/9510063

- Jaspers, Karl (1964), *Three Essays: Leonardo, Descartes, Max Weber.* New York: Harcourt, Brace and World.

- Jitatmananda, Swami (2006). *Modern Physics and Vedanta.* Mumbai: Bharatiya Vidya Bhavan.

- Kafatos, Menas, and Robert Nadeau (2000). *The Conscious Universe.* Berlin: Springer Verlag.

- Kahn, Charles H. (1985). *Anaximander and the Origins of Greek Cosmology.* Philadelphia, PA: Centrum Philadelphia.

- Lakshmanjoo, Swami (2007). *Kashmir Shaivism: The Secret Supreme.* Universal Shaiva Fellowship.

- Lanza, Robert (2009). *Biocentrism: How Life and Consciousness Are the Keys to Understanding the True Nature of the Universe.* Dallas, TX: BenBella Books.

- Laszlo, Ervin (1987). *L'Ipotesi del Campo Psi: Fisica e Metafisica dell'Evoluzione.* Bergamo: Pierluigi Lubrina.

- Laszlo, Ervin (1987b). *Evolution, The Grand Synthesis.* Boston, MA: Shambala.

- Laszlo, Ervin (1993). *The Creative Cosmos: A Unified Science of Matter, Life, and MInd.* Edinburgh: Floris Books.

- Laszlo, Ervin (1995). *The Interconnected Universe: Conceptual Foundations of Transdisciplinary Unified Theory.* Singapore: World Scientific.

- Laszlo, Ervin (1996). *The Whispering Pond: A Personal Guide to the Emerging Vision of Science.* London: Element Books.

- Laszlo, Ervin (2003). *The Connectivity Hypothesis: Foundations of an Integral Science of Quantum, Cosmos, Life, and Consciousness.* xxx: State University of New York Press.

- Laszlo, Ervin (2004). *Science and the Akashic Field: An Integral Theory of Everything.* Inner Traditions International, 2004

- Laszlo, Ervin (2004b). Systems Movement: Autobiographical Retrospectives, Ervin Laszlo. *Int. J. General Systems*, Vol. 33(1); pp. 1-14.

- Laszlo, Ervin (2006). *Science and the Reenchantment of the Cosmos: The Rise of the Integral Vision of Reality.* Rochester VT: Inner Traditions, 2006.

- Laszlo, Ervin (2008). *Quantum Shift in the Global Brain: How the New Scientific Reality Can Change Us and Our World.* Rochester VT: Inner Traditions.

- Laszlo, Ervin (2009). *The Akashic Experience: Science and the Cosmic Memory Field.* Rochester VT: Inner Traditions.

- Laszlo, Ervin, and Jude Currivan (2008). *Cosmos: A Co-Creators Guide to the Whole World.* New York: Hay House.

- Leibniz, Freiherr von Gottfried Wilhelm (16xx/2007). *Theodicy.* Charleston, SC: Bibliobazaar.

- Madore, J. (1999). *An Introduction to Noncommutative Differential Geometry and its Physical Applications*, 2nd edition. Cambridge: Cambridge University Press.

- Majumdar, Pradip Kr. (2002). The Buddha atomism (Internet resources).

- Mayer, Elisabeth Lloyd (2007). *Extraordinary Knowing: Science, Skepticism, and the Inexplicable Powers of the Human Mind.* New York: Bantam.

- McEvilley, Thomas (2002). *The Shape of Ancient Thought: Comparative Studies in Greek and Indian Philosophies.* New York: Allworth Press.

- McTaggart, Lynne (2001, 2008). *The Field: The Quest for the Secret Force of the Universe.* New York: HarperCollins, 2008.

- McTaggart, Lynne (2007, 2008). *The Intention Experiment: Using your Thoughts to Change your Life and the World.* New York: Free Press.

- Merleau-Ponty, Maurice (1968) *The Visible and the Invisible.* Evanston IL: Northwestern University Press. [Original French edition, Paris: Editions Gallimard, 1964]

- Moore, Thomas (1992). *Care of the Soul: A Guide for Cultivating Depth and Sacredness in Everyday Life.* New York: Harper Collins.

- Morwood, James, and John Taylor, eds. (2002). *The Pocket Oxford Classical Greek Dictionary.* Oxford: Oxford University Press.

- Odier, Daniel (2005). *Yoga Spandakarika: The Sacred Texts at the Origins of Tantra.* Rochester, VT: Inner Traditions.

- Onians, Richard Braxton (1951/1973). *The Origins of European Thought.* New York: Arno Press.

- Panda, N. C. (1995). *The Vibrating Universe.* Delhi: Motilal Banarsidass.

- Pandey, K. C. (2006). *Abhinavagupta: An Historical and Philosophical Study.* Varanasi, India: Chowkhamba.

- Pemmeraju, Sriram, and Steven Skiena (2003). *Computational Discrete Mathematics: Combinatorics and Graph Theory wth Mathematica.* Cambridge: Cambrige University Press.

- Penrose, Roger (1989). *The Emperor's New MInd.* New York: Oxford University Press.

- Penrose, Roger (1994). *Shadows of the MInd.* New York: Oxford University Press.

- Popper, Karl R. (1998), *The World of Parmenides: Essays on the Presocratic Enlightenment.* London: Routledge.

- Prabhananda, Swami (2003). *Philosophy and Science: An Exploratory Approach to Consciousness.* Kolkata, India: Ramakrishna Mission Institute of Culture.

- Prabhananda, Swami (2004). *Life, Mind, and Consciousness.* Kolkata, India: Ramakrishna Mission Institute of Culture.

- Prabhananda, Swami (2006). *Consciousness: A Deeper Scientific Search.* Kolkata, India: Ramakrishna Mission Institute of Culture.

- Prabhananda, Swami (2009). *Understanding Consciousness: Recent Advances.* Kolkata, India: Ramakrishna Mission Institute of Culture.

- Prasad, Rama (1889). *Nature's Finer Forces.* Adyar, Madras, India: Theosophical Publishing House.

- Proclus (1987). *A Commentary on the First Book of Euclid's Elements,* transl. Glenn R. Morrow. Princeton, NJ: Princeton Univ. Press,.

- Proclus (1970). *Commentary on Plato's Parmenides,* transl. Glenn R. Morrow and John M. Dillon. Princeton, NJ: Princeton Univ. Press,.

- Purton, W. J. (1890). *Pronunciation of Ancient Greek.* Cambridge: Cambridge University Press.

- Putterman, S. J. (1995). Sonoluminescence: Sound into light. *Scientific American* 272 (2): pp. 46-51.

- Radin, Dean (1997). *The Conscious Universe The Scientific Truth of Psychic Phenomena.* New York: HarperEdge.

- Radin, Dean (2006). *Entangled MInds: Extrasensory Experiences in a Quantum Reality.* New York: Paraview.

- Redei, Miklos, and Michael Stoltzner, eds. (2001). *John von Neumann and the Foundations of Quantum Physics*. Amsterdam: Kluwer.

- Requardt, Manfred, and Sisir Roy (2001). (Quantum) space-time as a statistical geometry of fuzzy lumps and the connection with random metric spaces, *Class.Quantum.Grav.* **18**, 3039.

- Ridondi, Pietro (1987), *Galileo Heretic.* transl. by Raymond Rosenthal. Princeton NJ: Princeton University Press.

- Rosen, Steven M. (2004). *Dimensions of Apeiron: A Topological Phenomenology of Space, Time, and Individuation*. Amsterdam: Value Inquiry.

- Roy, Sisir (1998). *Statistical Geometry and Applications to Microphysics and Cosmology*. Dordrecht; Boston: Kluwer Academic.

- Roy, Sisir (2003). *Quantum information and levels of consciousness*. In: Prabhananda, 2003; pp. 223-241.

- Rubenstein, Richard E. (2003). *Aristotle's Children: How Christians, Muslims, and Jews Rediscovered Ancient Wisdom and Illuminated the Dark Ages*. New York: Harcourt.

- Saraswati, Swami Satyananda (1998). *Yoga Nidra*. Munger, Bihar, India: Yoga Publications, Bihar School of Yoga.

- Saraswati, Swami Satyananda (2002), *Four Chapters on Freedom*. Munger, Bihar, India: Yoga Publications, Bihar School of Yoga.

- Saraswati, Swami Satyasangananda (1984). *Tattwa Shuddhi: The Tantric Practice of Inner Purification*. Munger, Biha, India: Yoga Publications Trust.

- Saraswati, Swami Satysangananda (2003). *Sri Vijnana Bhairava Tantra, The Ascent*. Munger, Bihar, India: Yoga Publications, Bihar School of Yoga.

- Saraswati, Swami Satyasangananda (2008). *Sri Saudarya Lahari*. Munger, Bihar, India: Yoga Publications Trust.

- Searle, John R. (1997). *The Mystery of Consciousness*. New York, NY: The New York Review of Books.

- Seifer, Marc (2008). *Transcending the Speed of Light: Consciousness, Quantum Physics, and the Fifth Dimension*. Rochester, VT: Inner Traditions.

- SenSharma, Deba Brata (1990). *The Philosophy of Sadhana, with Special Reference to the Trika School of Kashmir.*. Albany, NY: SUNY Press.

- SenSharma, Deba Brata (2003). Consciousness in Indian Philosophical Thought with Special Reference to the Advaita Saiva School of Kashmir. In: Prabhananda, 2003; pp. 207-222.

- SenSharma, Deba Brata (2004). Concept of Prana in Kashmir Saivism. In: Prabhananda, 2004; pp. 515-517.

- SenSharma, Deba Brata (2007). *Aspects of Tantra Yoga*. Varanasi, India: Indica.

- Serres, Michel (1968). *Le Système de Leibniz et ses Modèles Mathématiques*. Paris: Presses Universitaires.

- Shankarananda, Swami (2003). *The Yoga of Kashmir Shaivism*. Delhi: Motilal Banarsidass.

- Sharma, Subhash C. (2004). Vaisesika or the philosophy of Atomistic Pluralism (Internet resources).

- Shea, William R. (1991). *The Magic of Numbers and Motion: The Scientific Career of Rene Descartes*. Canton, MA: Science History Publications.

- Sheldrake, Rupert (1981, 1985). *A New Science of Life: The Hypothesis of Formative Causation*. London: Anthony Blond.

- Sheldrake, Rupert (1988). *The Presence of the Past: Morphic Resonance & the Habits of Nature.* New York: Times Books.

- Sheldrake, Rupert (1999). *The Sense of Being Stared At: And other Aspects of the Extended MInd.* New York: Crown,

- Sheldrake, Rupert (2003). *Dogs That Know When Their Owners Are Coming Home And Other Unexplained Powers of Animals: An Investigation.* London: Hutchinson.

- Singh, Jaideva (1980). *Spanda-karikas: The Divine Creative Pulsation.* Delhi, India: Motilal Benarsidass.

- Smolin, Lee (1998). The future of spin networks, in: S.A. Hugget et al (eds), *The Geometrical Universe.* Oxford : Oxford University Press.

- Smolin, Lee (2001). *Three Roads to Quantum Gravity.* New York, NY: Basic Books.

- Smolin, Lee (2006). *The Trouble with Physics.* New York: Houghton-Mifflin.

- Snell, Bruno (1960). *The Discovery of the Mind: the Greek Origins of European Thought .* New York: Harper.

- Sobel, Jyoti and Prem (1984/2007). *The Hierarchy of Minds: The MInd Levels.* Puducherry, India: Sri Aurobindo Ashram.

- Stapp, Henry P. (1993/2004/2009). *Mind, Matter and Quantum Mechanics.* Berlin: Springer Verlag.

- Stapp, Henry P. (2007). *Mindful Universe: Quantum Mechanics and the Participating Observer.* Berlin: Springer Verlag.

- Stapp, Henry P. (2009b). The role of human beings in the quantum universe. *World Futures* vol. 65 n. 1 (January 2009), pp. 7-18.

- Steiner, Rudolph (1909/1969). *Occult science - An Outline*. Trans. George and Mary Adams. London: Rudolf Steiner Press.

- Tart, Charles (2009). *The End of Materialism: How Evidence of the Paranormal Is Bringing Science and Spirit Together*. Oakland, CA: New Harbinger.

- Teilhard de Chardin, Pierre (1955, 1959). *The Phenomenon of Man*. New York: Harper & Row

- Thompson, Evan T. (2007). *Mind in Life: Biology, Phenomenology, and the Sciences of Mind*. Cambridge, MA: Harvard University Press.

- Thompson, William Irwin (1981). *The Time Fallings Bodies Take To Light: Mythology, Sexuality and the Origins of Culture*. New York, NY: St. Martin's Press.

- Tiller, William A. (2007). *Psychoenergetic Science: A Second Copernican-Scale Revolution*. Walnut Creek, CA: Pavior.

- Tillich, Paul (1963). *Systemic Theology, Volume III: Life and the Spirit, History and the Kingdom of God*. Chcago, IL: University of Chicago Press.

- Varela, Francisco J., Evan T. Thompson, and Eleanor Rosch (1992). *The Embodied Mind: Cognitive Science and Human Experience*. Cambridge, MA: MIT Press.

- Varela, Francisco J., and Jonathan Shear (1999). *The View from Within: First-person Approaches to the Study of Consciousness*. London: Imprint Academic. sw

- Von Neumann, John (1932/1955). *Mathematical Foundations of Quantum Mechanics*. Princeton: Princeton University Press.

- Walker, Evan Harris (2000). *The Physics of Consciousness: Quantum Minds and the Meaning of Life.* New York: Basic Books.

- Wheeler, John Archibald, and Wojciech Hubert Zurek, eds. (1983). *Quantum Theory and Measurement.* Princeton: Princeton University Press.

- Whitney, John H. (1980). *Digital Harmony,:On the Complementarity of Music and Visual Art.* Peterborough, NH: Byte Books.

- Wiener, Philip P., ed. (1951). *Leibniz Selections.* New York: Scribner's.

- Wigner, Eugene P. Remarks on the mind-body question (1961). In: Wheeler, 1983; p. 168. The problem of measurement (1963). In: Wheeler, 1983; p. 324. Interpretation of quantum mechanics (1976). In: Wheeler, 1983; p. 260.

- Woit, Peter (2006). *Not Even Wrong: The Failure of String Theory and the Search for Unity in Physical Law.* New York: Basic Books.

- Wolf, Fred Alan (1981). *Taking the Quantum Leap: the New Physics for Nonscientists.* San Francisco, CA: Harper & Row.

- Wolf, Fred Alan (1984). *Star Wave: Mind, Consciousness, and Quantum Physics.* New York, NY: Macmillan.

- Wolf, Fred Alan (1994). *The Dreaming Universe: a Mind-expanding Journey into the Realm where Psyche and Physics Meet.* New York, NY: Simon & Schuster.

- Wolf, Fred Alan (1996). *The Spiritual Universe: How Quantum Physics Proves the Existence of the Soul.* New York, NY: Simon & Schuster.

- Wolfram, Stephen (2002), *A New Kind of Science*. Champaign, IL: Wolfram Media.

- Yogananda, Paramahansa (2007). *The Bhagavad Gita*. Dakhneswar, India: Yogananda Satsang Society.

- Yukawa, H. (1966). Atomistic and the divisibility of space and time. *Supplements of the Progress of Theoretical Physics*, Nos. 37 & 38 (1966), p. 512.

Guide to Pronunciation

Greek

There has been significant progress in the reading of ancient Greek recently.[138] But to maintain simplicity, we will refer to this short list of Greek vowels taken from (Purton, 1890; p. vi).

ā as in father
a as in man
ī as in quinine
i as in quinine
ē as in fête
e as in ebb
ō as in note
o as in not
ū as in lute
u as in put

---

[138] See for example, (Daitz, 1981).

Figure G1. Romanization of Sanskrit Syllabary..

Sanskrit

Sanskrit has more syllables than English, and is usually transliterated with various diacritical marks. For example, there are three "s" sounds, and the word "akasha" in our title (pronounced ah-kash) is spelled "akasa" throughout our book. We will follow the most widely used romanization scheme for Sanskrit dictionaries, from the National Library of Kolkata, shoown in Figure 0.1. [139]

---

[139]en.wikipedia.org/wiki/National_Library_at_Kolkata_romanization

For pronunciation, it will be helpful to refer to these examples.[140]

a as in america, a
ā as in father, aa
i as in fill, i
ī as in police, ii
u as in full, u
ū as in rude, uu
ṛi as in merrily, ri
ṛī as in marine, rii
e as in prey, e
ai as in aisle, ai
o as in stone, o
au as ou in house, au
ś as in sure, sh
ṣ as sh in shun, sh
s as in saint, s

Glossary

Greek Words

Transliteration, [alternate spelling], meanings.[141]

apeiron, unlimited, boundless
arkhē, [arche], first principle
atomos, indivisible, uncut
To En, The One, monad, unity

---

[140] Adapted from (Lakshmanjoo, 1985).
[141] Adapted from (Morwood, 2002).

upostasos, [hypostasis], being
logos, saying, word, principle, rational basis
noos, [nous], mind, reason, thought, being, intelligence
physis, origin, nature, the universe, natural order
pneuma, wind, breath, life, spirit, mind, spiritual being
psukhē, [psyche], breath, spirit, life, living being, soul, desire, courage

Sanskrit Words

advaita, (alt aduita); not two, nonduality
ākāśa, (alt. akasha, akasa); ether, space
cakra, (alt. chakra); energy center
darśana, (alt. darshan); sight, vision
kośa, (alt. kosha); sheath, subtle body
māyā, (alt. maya); the principle of illusion
śakti, (alt. shakti); energy, potency
Śakti, (alt. Shakti); the spouse of Śiva
śiva, (alt. shiva); auspicious
Śiva, (alt. Shiva); the Ultimate Reality
spanda, pulse, vibration
tanmātras, the pure elements
tantra, ritual, religious treatise
trika, triple, Kashmiri Shaivism
tattva, category
vedānta, end of the Vedas, wisdom
yoga, yoke, path to oneness

Index

Abhinavagupta, 28
Advaita Vedanta, 31
Akasha tattva, 3, 35, 37, 64
al-Kindi, 14
Anaximander, 1, 7
Aphorisms of Siva, 32
Aristotle, 12
astrological magic, 16
atomism, 41

Bacon, 15

Capra, 63
cliques, 35
collapse of the wave function, 58, 61
condensation, 112
consciousness, 1, 22
cosmic consciousness, 1, 9
cosmic time, 113

Democritus, 42
Descartes, 18, 111
Dharmakirti, 45
digital philosophy, 3
Dirac, 51
discrete space and time, 117
dynamical cellular network, 25, 31, 48, 119

entanglement, 3, 67, 75

ether, 35, 37

Ficino, 16
Fredkin, 54
fuzzy lumps, 35

Galileo, 18, 47
Gilbert, 17
Goswami, 63

Heraclitus, 9
Hesiod, 7
Homer, 7
hypostases, 13

Ibn al-'Arabi, 15
idea, 9, 42
individual soul, 5, 11, 20
integral monism, 31

kancukas, 29, 35
Kashmiri Shaivism, 1, 26, 118
Kepler, 17

Laszlo, 64
Leibniz, 19
logos, 12

macroscopic spacetime, 113
mahabhutas, 30, 38
Maimonides, 46
materialism, 63

Maya, 35
measurement problem, 56, 58, 61, 75
meditation, 117
Merleau-Ponty, 21
meta-analysis, 68
microscopic time, 113
mind/body problem, 2, 10, 19, 31, 62, 67, 76, 109, 115
mind/matter interaction, 69
morphic fields, 22
morphic resonance, 22
morphogenetic field, 22
Motekallamin, 46

Nature, 21
Neoplatonism, 14, 15
nephesh, 7
nondualistic, 30, 118

orthodox interpretation, 61

paranormal phenomena, 2
Parmenides, 9, 41
Philebus, 12
Plato, 10
Plotinus, 13
positrons, 52
precognition, 72
presentiment, 70
primary qualities, 47
Proclus, 13
psi field, 73, 74
psyche, 7
Pythagoras, 8

quantum fields, 1
quantum foam, 60
quantum gravity, 53

quantum revolution, 49
quantum vacuum, 2, 50, 67, 117

Radin, 68
reincarnation, 6, 9
Republic, 11
ruach, 7

saccidananda, 31
Sacraments, 47
samvid, 31
secondary qualities, 47
Sheldrake, 21, 72
Siva sutra, 32
Siva tattva, 64, 117
Socrates, 10
soul, 7, 9
space, 27, 35
Spanda, 1, 27, 32, 33
Spandana, 33
Spirit, 7, 16, 21
Stapp, 57
Stoics, 12
Suhrawardi, 14
synechism, 43

Tantraloka, 28
Tantras, 27
tattvas, 1, 28, 40
telepathy, 69
thymos, 7
Tillich, 22
Timaeus, 11
time, 27
Trika, 26, 28
two times, 35

Ulam, 53

INDEX

Vasubandhu, 45
Vernadsky, 20
von Neumann, 53, 56

Whitehead, 59
Wigner, 58
World Soul, 5, 10, 14, 15, 21, 22

zero-point energy, 67

www.ingramcontent.com/pod-product-compliance
Lightning Source LLC
Chambersburg PA
CBHW032042150426
43194CB00006B/391